METROPOLIS AND REGION IN TRANSITION

Roster of Major American Centers and Their Symbols used Throughout this Volume

	Boston	Bo
	New York	NY
	Philadelphia	Pa
	Baltimore	Ba
As of 1860	New Orleans	NO
	Cincinnati	Ci
	St. Louis	SL
	Chicago	Cg
	San Francisco	SF
Added by 1880	Pittsburgh	Pi
	Buffalo	Bu
	Cleveland	Cl
	Detroit	Dt
	Milwaukee	Ml
	Washington	Wa
Added by 1900	Minneapolis	Mp
	Kansas City	KC
Added by 1920	Los Angeles	LA
	Houston	Ho
	Dallas	Dl
	Seattle	Se
Added 1940–60	Miami	Mi

METROPOLIS AND REGION IN TRANSITION

BEVERLY DUNCAN
& STANLEY LIEBERSON

SAGE PUBLICATIONS / Beverly Hills, California

For information address:

SAGE PUBLICATIONS, INC.
275 South Beverly Drive
Beverly Hills, California 90212

Printed in the United States of America

International Standard Book Number 0-8039-0051-1

Library of Congress Catalog Card No. 78-92350

First Printing

ACKNOWLEDGMENTS

The studies reported in Chapters 6, 11, 12, 16, 17, and 18 were carried out under the direction of Stanley Lieberson at the University of Wisconsin, Madison, during 1965–1967. The clerical assistance of Deborah Kuhn Meyersohn, David Sorenson, Ozzie L. Edwards, and Leslie G. Ibach and the secretarial assistance of Janice Deneen are gratefully acknowledged.

The other materials were prepared under the direction of Beverly Duncan at the Population Studies Center of the University of Michigan, with financial support provided by the Social Science Research Council during 1962–1963 and Resources for the Future during 1965–1967. We acknowledge with gratitude the assistance of: J. Michael Coble in programming; Ruthe C. Sweet in data processing; Elliot Long and Neil Paterson in compiling statistics; Joanne Raymond and Barbara Sain in coding and key punching; and Cheryl Arney, Susan Bittner, and Alice Y. Sano in the preparation of the manuscript.

The assistance of Charles B. McVey who prepared the graphics at the Office of Population Research, University of Washington was indispensable.

We are especially indebted to Otis Dudley Duncan, University

5

of Michigan, Eric E. Lampard, University of Wisconsin at Madison, and Harvey S. Perloff, Resources for the Future, who read various draft documents and encouraged us to continue our work.

Beverly Duncan
Stanley Lieberson

Seattle and Ann Arbor
March 1969

CONTENTS

Contents

PART III

SPECIAL STUDIES

TABLES

APPENDIX

FIGURES

Chapter 1

BACKGROUND

Ten years ago a study of the metropolitan structure of the United States was undertaken to establish a "mid-century bench mark" which would both document the state to which the metropolitan economy had evolved by 1950 and provide a baseline for measuring subsequent change. The issues confronted by our collaborators and us in the study were the nature of the metropolis, its role in the national economy, and its relation to regional differentiation of the economy. We recognized that because our conclusions would rest on observations made at a single point in time, a transient conjunction of circumstances might easily be mistaken for a basic structural pattern. Nonetheless, we argued, adequate descriptions of present realities should precede their historical explanations as well as forecasts of developments to come.

Like any complex and polymorphic entity, the metropolis is subject to differing interpretations according to the perspective from which it is viewed; but whatever meaning of "metropolis" is accepted, the entity usually is conceived as a special kind of city. In a formulation of metropolitanism that has become classic, Gras asserted nearly a half-century ago that population size alone was too simplistic for identifying the metropolis. A relatively large population was required for, but did not ensure, metropolitan

17

status. The metropolis was to be industrially developed, but would have relatively few workers in manufacturing as compared with neighboring centers. The functional emphasis of his conception can be captured in the following quotation:

> that city is a full-fledged metropolis when most kinds of products of the district concentrate in it for trade as well as transit; when these products are paid for by wares that radiate from it; and when necessary financial transactions involved in this exchange are provided by it. [Gras, 1922: 294]

The Gras conception of the metropolis appears to embody both the central-place and the break-in-transportation principles of urban location and function (e.g., Harris and Ullman, 1945: 7–9), which would imply at least two characteristic metropolis-region relations. Ties would obtain between the center and the surrounding territory and also between the center and remote areas intersected by transport radials focused on the center. That Gras was sensitive to this duality perhaps can be accepted given his emphasis on the roles of the center as coordinator of both local and long-distance trade or as organizer of the surrounding territory and its link with the rest of the world.

One such form of metropolis-region relation seems to have been captured in the concept of the metropolitan region. When a so-called metropolitan region has been delimited, it typically is a more or less extended area continuous with the center and in which center-based economic units are "active" by some absolute standard or relative to units in other centers. There may also be the implication that the region is organized by the center, but regional forces, in turn, shape the course of development in the center.

Guided by notions of this sort, vague and sometimes contradictory, we began the study that was reported in a volume entitled *Metropolis and Region* (Duncan et al., 1960). When the study was completed, the formulations of metropolitan structure distilled from the literature seemed less cogent than they had at the outset. Not every large population center in the nation need be a "metropolis" nor need the "metropolitan region" be relevant to all the functions carried out in a center in order for these concepts to be powerful tools in understanding the spatial organization of the national economy. They should capture a significant part of the realities of the structure, however.

The salient feature of the metropolis has been identified as its sizable commercial-financial complex. When the local economies of the nation's larger urban centers were examined in some detail at mid-century, some larger places clearly met the criteria of strength in the commercial-financial sector. Such a metropolis might or might not have a relatively high level of manufacturing activity; in fact, the two possibilities appeared about equally probable. Other urban centers with equally large populations were observed in which manufacturing activities were prominent in the industrial profile, but in which commercial and financial functions were poorly developed. The fact that many large centers appeared not to qualify as a metropolis in no sense invalidates the concept of the metropolis, but it does mean that the concept of metropolis failed to account for many major population concentrations in the nation.

The metropolitan region has been conceived as a continuous hinterland that is organized by units located in the metropolis and includes important supply and market areas for the center's industrial specialties. In fact, the proportion of industrial specialties for which such a metropolitan region is relevant to the actual center-region relation was found to be astonishingly low. Supply and market areas for some industries fell within the center itself. A supply area for another specialty might be shared by centers located at variable, but considerable, distances from it; or the outputs of a center specialty might be destined for the national, rather than a regional, market. The form of metropolis-region relation captured in the concept of the metropolitan region did appear, but it did so relatively infrequently.

With observations on the spatial structure of the economy restricted to a single point in time, it was difficult to determine why the classic metropolitan formulation failed to reflect faithfully the realities of the mid-century structure. A situational factor seemed to account, at least in part, for the diversity of industrial profiles and regional relations exhibited by the large centers. Generalized accessibility to the national market appeared highly favorable to manufacturing specialization. In sections of the nation where such accessibility was high, large centers were founded to be closely spaced. There was inadequate "tributary territory" for the development of a commercial-financial specialization in each center. With the possible exception of a very few key centers like New York and Chicago whose tributary territory may be coextensive with the nation, no center in the so-called northeastern manufacturing belt evidenced close ties between its industrial specialties and a continuous hinterland. Some manufacturing-belt centers were found to have commercial and financial functions commensurate with their population size, but in a majority the distinctively metropolitan function was not prominent in the local economy. The classic metropolitan formulation offered a more adequate account of centers scattered outside the manufacturing belt, separated from one another by considerable distances.

Recently we have been trying to illuminate the contemporary structure by tracing its development in the context of the peopling of the continent and the industrialization of the national economy. Our description of the American "case history" of an evolving system of major urban centers is set forth chronologically. In *Metropolis and Region,* the recurring theme was "what a city does depends on where it is"; here we are guided by the notion that what a city does depends not only on where it is, but also on what it has done in the past.

Many readers will be more literate about the American history than we are; others will have a better command of the theory that bears on the spatial distribution of economic activity. Our only justification is a conviction that a historical perspective is needed to understand why cities now play distinctive roles in the spatial organization of the national economy.

Our interest was aroused by two unrelated and unsystematic observations. First, we had occasion to map the "age" of the nation's larger cities. The oldest group of cities clustered in the east, and successively younger groups typically fell further to the west. There was, however, a backtracking from trans-Mississippi territory into the trans-Appalachian north among cities "born" late in the 1800s. By one reckoning of age, those manufacturing-belt centers where the metropolitan function had been found to be most poorly developed were younger than the centers that encircled them. Second, we had occasion to inspect the industrial profiles of the "newest" centers that had been studied individually in

Metropolis and Region for possible common industrial specialties. The specialties appearing with relatively high frequency seemed to be in the newest lines of activity, most notably perhaps, air transportation and the manufacture of aircraft.

We present here a more systematic, though still somewhat unrelated, set of observations that bear on the relation between a center's growth history and its current economic structure.

Recounted first is the planting of mercantile outposts in untapped American resource regions on the Atlantic seaboard by colonizing companies based in European metropolitan centers; the regional specialties through which colonials extracted an exportable surplus, the return on the investment in colonization; the growth of town life and a business community at colonial seaports; the competition between colonials and European investors for the profits from a swelling colonial trade.

Continental colonialism replaces European colonialism. Financial groups headquartered in port communities on the settled seaboard vie to control the terms under which the interior will produce an exportable surplus and enter it into the long-distance trade flows. Established commercial centers on the seaboard become differentiated in scale and mix of commercial activity by the success of their businessmen in organizing new territory.

As the full continent is integrated into a metropolitan economy, the process repeats with appropriate change in the roster of settled regions and established centers, and with the addition of railroads to waterways as a practicable means of moving the exportable surplus. By the close of the nineteenth century, a network of outstanding commercial centers scattered across the nation had become discernible.

These commercial centers were not, however, the only sizable population centers in America. The American economy had been industrialized. At the founding of the nation, perhaps no more than a tenth of the national population derived a livelihood from commerce or the "manufactories." Most Americans were both agriculturalists and household manufacturers. A century later trade flows had increased to a point where commerce provided employment for about a sixth of the national work force. Husbandry and manufacturing had become distinct pursuits, with agriculture providing employment for something over a third of the American workers and factories providing jobs for about a fifth of the work force. With the industrialization of manufactures came a new kind of city.

From the "steam-and-steel" complex, there developed large-scale concentrations of manufacturing activity. The location of iron ore and coal deposits, accessibility to the national market for manufactured goods, the placement of transport and communication channels combined to the favor of the northeast. Concentrations of manufactures sometimes developed in the immediate environs of "metropolitan" centers. In other instances, they developed in the environs of lesser places within the metropolitan regions of the northeastern metropolises.

There had emerged by 1900 a territorial division of labor which extended into the ranks of the nation's largest cities. Centers whose national prominence rested exclusively on their role as metropolis for a resource region still were to be observed—New Orleans, San Francisco, or Minneapolis. Other metropolitan centers had enhanced their strategic roles in the national economy by gaining a sizable concentration of industrial manufactures—Boston, New York, Philadelphia, and Baltimore, Cincinnati, St. Louis, and Chicago. For still other centers, the rise to national prominence coincided with the development of a large-scale local manufacturing concentration—Pittsburgh, Buffalo, Cleveland, Detroit, and Milwaukee. In part in response to the need for regulation in an economy with so fine a territorial division of labor, federal activity swelled; and Washington joined the ranks of the major centers. It is the development of this system of centers that is traced in Part I of our report.

Added to this roster of major centers shortly after 1900 were Los Angeles and Kansas City, but the roster then remained unchanged until 1950. The timing of change in the roster is definitional, to be sure. Nonetheless, given our definition of a major center, the timing is substantively interesting. It is the case that until the 1860s, the nation's top five cities were seaports whose origins trace to the colonial period. The smallest of the port communities was New Orleans. First to displace New Orleans in rank with respect to population size were Cincinnati, St. Louis, and Chicago, cities of the interior occupying strategic positions on the internal waterways leading to territory recently opened to resource exploitation. In the 1870s, San Francisco—a gateway to the far west—displaced New Orleans; but so did Pittsburgh. Within two decades, Minneapolis—a gateway to the mid-continent plains—had displaced New Orleans; but so had Buffalo, Cleveland, Detroit, and Milwaukee, as well as Washington. It is our contention that the rapid succession of new centers breaking into the ranks of the established centers in the late 1800s reflects a continuation of "continental colonialism" coupled with a reorganization in the spatial structure of economic activity occasioned by the industrialization of manufactures.

We know of no guideline for evaluating whether change in the system, that is, in the roles played by the respective major centers, has proceeded rapidly or slowly since 1900, even assuming adequate measures of change. Here we can only evaluate change in role by comparing current measures of each center's activity in two areas of finance—correspondent banking and municipal-bond underwriting—and in the size and industrial composition of the manufacturing sector with 1900 benchmarks established in Part I.

It is our impression that change has been relatively slight. Such change as is observed in the financial sphere seems to trace primarily to the competitive success of "new" indigenous centers vis-à-vis "old" extraregional centers within sections of the nation where the character of the economy has been undergoing relatively rapid change. Change in the manufacturing sector seems to trace primarily to the fact that activity in "new" lines of industry is distributed territorially more or less independently of the "old" distribution of manufacturing activity.

Reorganization in the system of major centers may again be gaining momentum. Since World War II, Houston, Dallas, Seattle, and Miami displaced New Orleans in rank with respect to population size. Estimates of metropolitan populations made in the mid-1960s open the possibility that within a decade Atlanta, San Diego, or Denver may outrank New Orleans. In fact, San Bernardino, San Jose, and Phoenix would become contenders for status as major centers were their current growth rates to continue (*Current Population Reports*, Series P–25, No. 347).

After examining the 1950 industrial profiles and the location of supply and market areas for key local industries in some fifty metropolitan centers, Duncan et al. (1960: 259 ff.) sought to cap-

ture the primary bases of differentiation among centers with a
sevenfold classification. San Bernardino and San Jose were smaller
in population size than the centers studied individually. The other
centers gaining prominence in the past decade were: one "regional
capital, sub-metropolitan"; four "regional metropolises"; and three
"special cases."

Houston was identified as a regional capital, sub-metropolitan,
the class exemplified by New Orleans among the established centers. Dallas and Seattle, Atlanta and Denver were included in the
regional-metropolis class along with the established centers of San
Francisco, Minneapolis, and Kansas City. In such places,

> Trade functions are moderately to highly developed. Typically,
> though not without exception, there are profile manufacturing in-
> dustries depending for important inputs on resource-extracting
> activities in the hinterland or a broader but still contiguous
> "region." Trade areas of moderate to large size can be more or less
> realistically identified, and there is clear evidence of the role of the
> center in integrating activities of such areas. [Duncan et al., 1960:
> 269]

Miami was classed as a special case on the grounds that its
complement of functions was atypical for a large center. Its distinctive profile was thought to be partially attributable to "rapid
growth superimposed on the response to the tourist trade" by a
number of so-called local service industries (Duncan et al., 1960:
273). San Diego was among the centers that seemed "underde-
veloped industrially and commercially on the basis of their size,"
which was perhaps overstated by the inclusion of "major military
installations and associated activities" (Duncan et al., 1960: 273).
Phoenix was a center in which the use of the county as the basic
unit in the delimitation of metropolitan areas and the recent rapid
growth in the locality seemingly distorted the industrial profile.
An atypically large amount of extractive activity was occurring
within the metropolitan area, and construction and some so-called
local service industries were most likely passing specializations
associated with growth. Specializations of long-run importance may
have been masked by the growth-induced specializations (Duncan
et al., 1960: 273–274).

A counterpart to the steam-and-steel complex which fostered

the rise of a new kind of city nearly a century ago is not easily identified, perhaps because its effects only now are becoming evident. One salient feature of the mid-century setting is the airplane, which simultaneously changed the scale of distance between established centers and opened a new line of manufacture with a distinctive resource-market orientation. No less noteworthy are the increasing share of economic activity coordinated directly or indirectly by the federal government and the rising real income and growing leisure time of the American work force. Whether in retrospect these factors will form a complex accounting for the rise of the "special cases," we do not know.

We conclude Part II having documented only that by the time centers gain national prominence, they have developed a distinctive industrial character which tends to persist thereafter; and that the centers now gaining prominence differ functionally from those places that consolidated their positions as major centers some decades ago.

In a final section, we investigate the destinations of goods manufactured in the major centers in 1963, the commercial specialties of these centers in 1960, and the way in which correspondent-banking networks developed in Texas and Florida, respectively, after 1900. These studies are grouped together only because each is severely limited with respect to time span or areal scope. They are included because each explores in more depth a topic touched on in the main line of our story. We conclude this volume by sharing with the reader some impressions held by bankers about the contemporary metropolitan structure or, more specifically, the territorial organization of financial activity in America.

It has not been our intention here to summarize the mass of descriptive material that follows. Neither have we attempted to draw "conclusions" about the nature of the processes that result in the sort of transformation we describe. We only set forth some impressions which we hope will arouse the curiosity of other investigators who are better equipped to pursue the problem of "why cities do what they do."

PART I

DEVELOPMENT OF A SYSTEM OF MAJOR CENTERS

Chapter 2

THE EUROPEAN METROPOLIS & AMERICAN COLONIZATION

Modern man appeared in Europe perhaps 30,000 years ago. In the West European context from which American colonization arose, he lived for most of that time as a hunter and gatherer. No more than 5,000 years ago did Neolithic agriculture begin to displace hunting and gathering as the economic base of human communities in this area. The appearance of an enduring town-oriented economy organized through town-situated commercial institutions can be dated no more than 1,000 years in the past, and the beginnings of metropolitan organization can be traced back less than 400 years.

While the hunting-collecting way of life prevailed, effectively organized human communities were small and identified only loosely with a territorial unit. A score or two of men moved about through a hunting ground, feasting at the time of a kill, starving if they came upon no game. Although the hunting band was a self-sufficient economic unit, its members shared a precarious livelihood. Occasional contact and possibly sporadic exchange between neighboring communities can be presumed. As McNeill (1963: 18) has observed, however, "Communities of hunters, whose way of life was essentially uniform throughout wide areas, and whose skills were exquisitely adapted to the existing environment, could find little stimulus to social change from contact with neighboring

29

human groups." The hunting populations did not disappear as cultivators began to move into Europe from their Middle Eastern center of origin after 4,000 B.C.; but as cultivators encountered hunters, contact was established between communities with different ways of life.

The pace of change quickened as Neolithic agriculture diffused over Europe. A mix of animal husbandry and crop cultivation became established as the economic base of the human community. Identification of the community with a territorial unit, a village, became closer as the agriculturalists took up permanent occupancy. Heightened per capita productivity and a growth of arts and crafts were almost certainly outcomes of the new way of life. When a prototypic plow was hitched to an animal to break the ground for planting, an assured food surplus became a reality. Man had, for the first time, tapped an external source of energy and thereby raised his productivity. He now had "leisure time" during the slack periods in the seasonal cycle of field work, and surviving artifacts suggest that the villager became a potter and tool-maker or perhaps a wood or leather worker in his leisure hours.

Before the diffusion of Neolithic agriculture across Europe was complete, what Lampard (1965: 532 ff.) has called "the capacity for nodal organization" was manifest in the Middle East. On a foundation of village agriculture, far-reaching innovations in the organization of economic activity were occurring. Repeated contact between peoples with different ways of life led to recurring frictions as well as transactions among them. To mediate the emerging patterns of interdependency, a class of "managers" emerged; and kinship ultimately gave way to kingship as the organizing principle of human communities. The managers whose authority often was legitimated both militarily and religiously had a "capacity to generate, store, and utilize social saving" and their seat became "a focus of generalized nodality" (Lampard, 1965: 523, 540).

The early civilizations which arose in the Middle East probably had little direct effect on the West European hunters and villagers although men in touch with civilized culture reached Western Europe at a relatively early date. McNeill (1963: 100) infers from the distribution of megaliths that colonization along the coasts by a seafaring elite of priests and merchants was underway nearly 4,500 years ago. More than 3,500 years ago, barbarians acquainted with bronze metallurgy from contact with Middle-Easterners had carried their warrior culture overland to the same coasts (McNeill, 1963: 103); but it was not until some 1,000 years ago that Western Europe itself became a center of civilization.

In his monumental survey of *The Rise of the West,* McNeill (1963: 451 ff.) argues that the institutional arrangements which permitted the West Europeans to exploit technological advances and a set of potentially advantageous site factors took shape in a setting of "cultural stagnation, political disintegration, and economic backwardness." The setting may, indeed, have been favorable; for West Europeans, relatively unencumbered by civilized institutions, could use selectively the cultural inheritances of other peoples. McNeill singles out agricultural advance, military advance, and "the special role of piratical trade" in detailing the "extraordinary upsurge of medieval Europe."

Improvements in shipbuilding opened the northern seas to regular travel and simultaneously fostered piracy on a new scale. Raiding rendered land-ownership rights meaningless and opened the way for a pooling of resources and redistribution of land rights. Cooperative tillage of open fields with heavier plows and increased animal power could be practiced. As additional land came under cultivation, "medieval Europe attained an agricultural base broad enough to sustain both a numerous military aristocracy and a vigorous town life."

The military professionals, well equipped and trained, provided effective defense against the plunderers and, in time, "found outlet for their prowess in foreign conquest and adventure." In the face of effective local resistance, "pirate ships and raiding parties predictably gave way to merchant shipping and pack trains." From itinerant trading bands developed settled merchant communities which in turn became the core of commercial towns. And, as Gras (1922: 105) observed some years ago, the transition from village to town economy was marked not by the appearance of trade, but by the appearance of traders, specialists in exchange, clustered at strategic locations.

Supported by growing internal prosperity, West European merchants and militarists directed their energies to territorial expansion. Crusading brought unprecedented numbers of West Europeans in contact with Eastern civilization, set in motion new trade flows, and stimulated innovation in financing long-distance trade and travel. Soon sea lanes had led the West Europeans across the Atlantic, and shortly after 1600 the colonization of America was underway.

EUROPEAN METROPOLITANISM

A transition from town to metropolitan organization in Western Europe can be placed between 1600 and 1700, provided majority participation in a market economy, a quickening growth in trade centers, and rapid elaboration of commercial institutions are accepted as indicators. Both Buer (1926: 61) and McNeill (1963: 584) observe the spread of commercial agriculture, a shift from "communal and for subsistence" to "capitalistic and for a market"; and McNeill also comments upon the growing share of West Europeans purchasing commodities of distant origin as bulk goods were added to the ancient luxury trade. Commercial centers on which overseas, coastal, and overland trade routes converged, perhaps most notably London, experienced an influx of population unparalleled in the past. Gras (1922) attached special importance to the financial indicators of metropolitanism, among which banking and joint stock trading are singled out by Buer (1926: 48) as noteworthy.

The hinterland of the West European metropolis was being extended across the Atlantic to encompass untapped resource areas, the yield of which could further swell the trade flows. Bringing the North American mainland into a European-centered metropolitan organization entailed no costly clash of civilizations or cumbersome grafting of a metropolitan-oriented sector to a village economy. The course of development in a territory so thinly peopled, little exploited, and loosely integrated could be set in the metropolitan centers. The frictions in colonization that posed "managerial" problems arose within the metropolitan business community more often than between its members and native Americans, at least until the American-born offspring of European settlers became business competitors.

By choice or necessity, European-based companies differed in their American investment strategy. The earliest settlements were planted by companies whose primary interest was exclusive control of the terms under which commodities of American origin entered the market. Short-run profits should peak if an indigenous product, valuable in the European market and extractable without much modification of the local economy, were to be found in abundance. A scarcity of such products gave rise to other forms of exploitive relation between the European metropolitan centers and the colonial resource regions, notably the plantation system and the overseas "European-style" community. (The distinction among forms of exploitive relation follows McNeill, 1963: 656.)

Colonization of the American mainland became a reality in 1607 when the London Company, "certain Knights, Gentlemen, Merchants, and other Adventurers of our city of London and elsewhere" (Faulkner, 1943: 48), planted a settlement thirty miles up the James River in Virginia. The charter under which the joint-stock company operated granted monopolistic trade rights with colonization privileges. From the outset, the venture was unprofitable. Gold and silver were not to be found near the James, and the shiploads of timber sent back to England failed even to meet the expenses of provisioning company workers. Reorganization to raise additional capital recurred as the Virginia Plantation struggled for survival. Not until it was found that local soil and climate were suited to the growing of a crop which was in demand in the European market and could not be supplied by the more immediate hinterland did the colony prosper. Prosperity was tied, however, to the production of and demand for the single crop, tobacco. A plantation form of enterprise, based on forced labor and large land holdings, took shape, which was to influence regional economic activity well beyond the colonial period.

Less than ten years after the James River settlement was established, Amsterdam merchants set up a trading post at the mouth of the Hudson River. The Dutch West India Company, a "corporation giant" of the day, acquired the Hudson Valley with its fur trade as one of the company's lesser overseas interests in the following decade. Company workers were scattered through the area at trading posts and clustered at the strategic transit points, New Amsterdam at the mouth of the Hudson and Fort Orange upstream at the head of navigation for ocean-going vessels. Trade in furs was a profitable, but transitory economic specialization. Its

long-run significance lay in opening the western frontier to exploration, not in any impact on the Hudson Valley economy. A
more direct influence on the future course of regional economic
activity came from the company's "members' option." Although the
company as such did not undertake large-scale settlement, company members had an option: tracts of land with sixteen miles
of river frontage and proprietary rights thereto for the transport
and settlement of fifty persons on the land.

Permanent settlements were established on the coast of what
now is Massachusetts a few years after the Hudson Valley outpost
had been established. The impetus for colonization in the case of
the Plymouth Plantation came from the prospective settlers, religious nonconformists in their home society. To gain the necessary
financing, they had to ally themselves with London merchants in a
colonizing company. The venture was unprofitable as an investment, for the abundant, valuable, easily extracted, indigenous product was missing. Claims of the London stockholders on the colony
ultimately were bought from meager profits of the fur trade carried
on by settlers in the transplanted "European community." Control
of the Massachusetts Bay Company, in contrast, resided with the
settlers from the outset. The company's charter had been obtained
in England for commercial purposes by a group which included
"Puritan" merchants. Royal policy with respect to religious freedom
became increasingly repressive, and in 1630 company "headquarters" were relocated in America.

These sketches (based on Faulkner, 1943: ch. 3) serve to illustrate the metropolitan organization through which American
colonization was effected. The origins of our nation's oldest cities
do not lie in a village or town economy. They were in the beginning
mercantile outposts in a resource region organized through the
developing metropolitan centers of Western Europe.

The largest nucleated settlements in the colonies were "villages"
and the vast majority of the colonists were "agriculturalists"; but
colonial America was part of a metropolitan economy centered on
Western Europe. Although no settlement had reached the thousand
mark in population by the mid-1630s, their residents were uncommon villagers by background. All were newcomers who had
traveled far. Included among them were men experienced in an
urban way of life, men schooled in the cultural inheritance of the

West, practiced in the conduct of long-distance trade, trained to high standards in their crafts. Most became at least part-time agriculturalists, for food was a pressing need. Yet, agriculture too had a rather special meaning: its products were leading items of export, a major source of wealth. When an opportunity arose, the colonists were quick to participate in the metropolitan market.

Differences among colonies in economic activity were developing before settlement of the Atlantic seaboard was complete. Economic specialization could occur because the colonial resource regions were part of a much larger organization of producers and consumers, and the forms such specialization took can be understood only in the context of the larger organization. The "north-south" difference in economic activity along the coast sharpened, for example, as specialization in a neighboring colonial resource region became more pronounced. In the Caribbean, a plantation system built around the production of sugar displaced tobacco growing and the cultivation of food crops. European demand for American tobacco, the leading southern export, became greater at the same time that a Caribbean outlet was opened for fish and grains, export commodities of the northerly colonies. The heightened specialization implied growing economic dependency not only between colonies and the European core, but also among colonial resource regions.

Businessmen in the mother country pressed for regulation of both trans-Atlantic and intercolonial trade as their overseas investment grew and shipments bound from or for colonial ports became more numerous. The regulatory policies instigated to prevent diversion of their profits to European competitors carried an occasional side-benefit for the American colonials. The stipulation that ships engaged in the trade must be British-built, for example, was to the advantage of shipbuilders in New England, where the supply of timber along the coast was more than sufficient to meet the locally generated demand for fishing boats. More often, however, the restrictions worked a hardship on American colonials or would have done so had enforcement been effective. The monopoly for English middlemen established by controls over long-distance shipping routes is a case in point. Colonial exports destined for the trans-Atlantic market were to be shipped solely to England for use in the mother country or for re-export to other parts of the market; colonial imports from European supply areas were to originate in

England or to be shipped by way of the mother country. (A fuller description appears in Faulkner, 1943: 113–125.)

Such regulatory policies were intended to protect the economic basis of the developing metropolitanism in the mother country. Materials from the colonial resource regions were to supplement, not supplant, locally available materials. Colonial populations were to offer an assured market for any surplus output of local industries. Profits from colonial trade were to accumulate in the metropolitan centers, collecting and distributing sites for long-distance trade.

In earlier decades, an assured agricultural surplus at home had supported the overseas ventures of merchants and militarists. The metropolitan centers now provided agriculture in the surrounding area with both a market and capital sufficient to effect industrialization in the agriculture sector. (Buer, 1926: 47 ff., describes the interplay of commerce and agriculture.) Before the eighteenth century closed, the semi-rural towns in which small-shop manufactures clustered "were becoming a major outlet for capital accumulated in commercialized agricultural production," according to Lampard (1955: 105). The industrialization of the manufacturing sector was underway.

The outlines of an industrial economy organized through metropolitan centers were, then, visible in England by the time the American colonies united to form a new nation. The transformation of a town-oriented, nonindustrial economy was documented in the English experience; and the impact of metropolitan reorganization on the spatial structure of a national economy could be measured grossly by the redistribution of population and economic activity. The English experience does not offer much by way of guidelines for tracing the development of the American economy, however. The American starting point for metropolitan development was a largely unsettled and unexploited resource region with economic ties to an overseas metropolitan core.

THE COLONIAL RESOURCE REGIONS

Fewer than 5,000 Europeans were living in the American colonies along the Atlantic seaboard by 1630. About 2,500 lived in the Virginia colony, for the most part in scattered settlements along the James River. To the north, separated by hundreds of miles, were about 2,000 settlers, their villages strung along the Atlantic from the lower Maine coast to Cape Cod. Between were a few hundred Europeans in what was to become the New York colony.

A census taken in 1720 would have enumerated about 400,000 settlers of European origin and some 70,000 Negroes who had been imported to meet a colonial labor shortage. The population of the younger colonies nearly outnumbered residents of Virginia and Massachusetts. The seaboard was filling in rapidly with settlements of whites, and a concentration of blacks was only slightly more pronounced in the southern plantation districts than elsewhere (Figure 2-1).

In contemporary economic parlance, the colonial industrial profile of 1720 was weighted heavily by "primary resource extractors," agriculture, forest industries, fishing. Two crops of the plantation system were being shipped out in substantial volume: exports of tobacco from Virginia and Maryland to England had reached half their maximum for the colonial period; and exports of rice from the Carolinas and Georgia had reached a tenth their maximum. The value of furs shipped from New England and New York to England already was near its peak, and dried fish were still another important export commodity of the north. The importance of corn and wheat, bread and flour, beef and pork, or barrel staves, pitch, and tar in colonial trade channels can be inferred from their inclusion in published lists of "prices current." As revealed by the fragmentary statistics on the colonial economy, the industrial profile was rounded out by a few "first-stage resource users," grain and lumber milling and meat packing, and an occasional "second-stage resource user," woodworking or baking.

There was, as yet, little "town life" in the colonies. Exports of the profitable southern plantation system tended to move directly from producer to the English market, damping the development of local mercantile institutions or town-based industries. Resource extraction also dominated the northern industrial profile, but the mix of economic activity and absence of an assured English market were more favorable to the growth of commercial centers. The larger nucleated settlements, none of which probably had more than 10,000 residents, had grown up around the northern ports of Boston, New York, and Philadelphia.

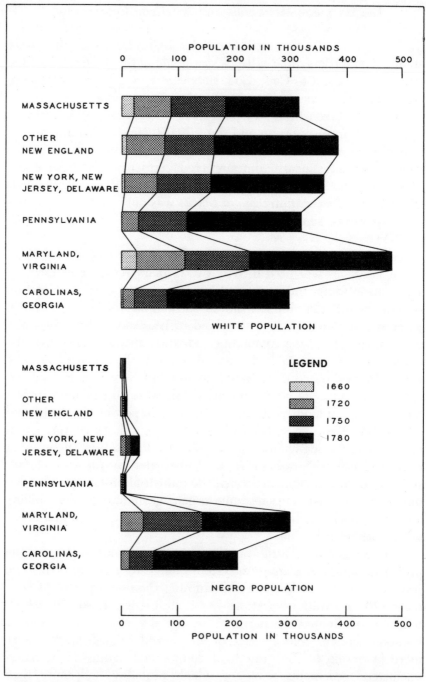

Figure 2-1. Population, by color, of American colonies: 1660 to 1780. (Source: *Historical Statistics of the United States,* Series Z2–17.)

The colonial population of European origin continued to grow rapidly after 1720, doubling in size every thirty years. Importation of Negroes by the plantation districts accelerated so that the growth rate for the blacks was even more rapid than the rate for whites. The northern port communities of Boston, New York, and Philadelphia maintained their positions as the largest nucleated settlements in the colonies, and by the 1770s their populations were approaching the 20,000 mark. To the south only the settlement at the port of Charleston on the South Carolina coast had reached the 10,000 mark in population size.

Increases in shipping activity at the ports during the 1700s are a matter of record. The growing volume of shipments passing through northern ports reflected primarily expansion in trade among colonies; to the south, the gross flow with England had swelled (Figure 2-2). The economic tie of colonial producers and traders to the English market loosened only in the north, where the exports of leading value were foodstuffs entered into "triangular" trades through colonial plantation districts or southern Europe to balance the necessary imports from England. Southerners were contributing disproportionately to the total value of exports from the mainland colonies, and they did so as suppliers of tobacco, rice, and indigo to the English (Figure 2-3).

The continuous area opened to settlement by Europeans still extended inland only a hundred miles or so along most of the Atlantic coast, but isolated settlements had been established west of the Appalachian Mountain barrier. As spillover of settlers through mountain passes into the interior began to gain momentum, colonial land policy became as salient an issue as colonial trade policy in the mother country. A century earlier, English investors had been challenged by other European businessmen in organizing the resource regions along the Atlantic coast. The challenge in organizing the interior resource regions came from American colonials. In response to the new challenge, the mother country imposed restrictions on land acquisitions by colonials.

In the years between 1770 and 1810, the political independence of the colonies from the mother country was declared; the Constitution of the United States was adopted by the former colonies; and the territorial extent of the new nation was doubled by the Louisiana Purchase.

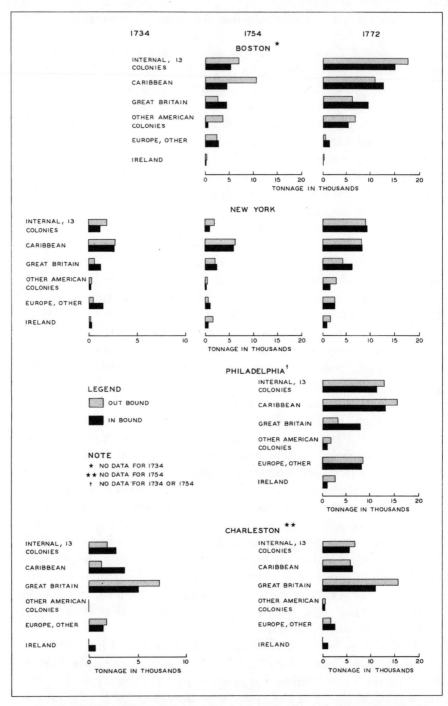

Figure 2-2. Leading destinations and origins of ships outward and inward bound, for selected colonial ports in the 1700s. (Source: *Historical Statistics of the United States,* Series Z56–75.)

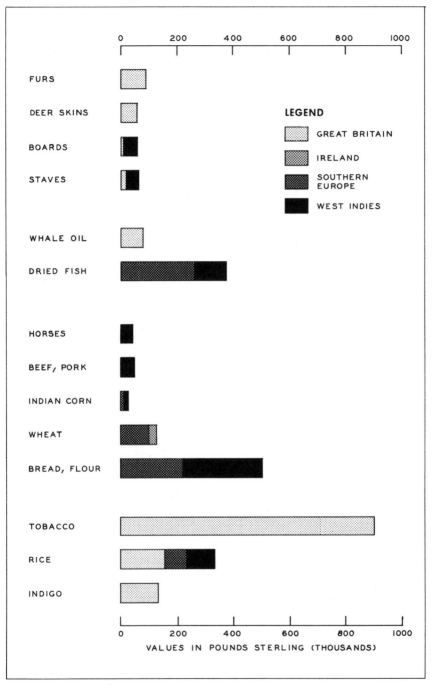

Figure 2-3. Exports of leading value, by destination, for British continental colonies: 1770. (Source: *Historical Statistics of the United States,* Series Z76.)

The Union remained heavily dependent on overseas markets and supply areas, exported predominantly crude materials, imported predominantly finished manufactures, and traded with the United Kingdom more than with any other nation. As colonials Americans had manned ships built in their own colonies in the long-distance trade and had located markets for their commodities which competed with, rather than complemented, English products. Acts of the First Congress of the United States provided for the compilation of records on foreign trade and protected American shipping interests against foreign competition.

With political independence, the Americans controlled land policy on the western frontier. Before 1810 four inland states had been admitted to the Union: Vermont, Kentucky, Tennessee, and Ohio. Population in the new states exceeded a million, accounting for a seventh of the 1810 national total. Another 75,000 persons were living in trans-Appalachian territory not yet admitted to statehood, but settled densely enough to be covered by the census enumeration of that year; and 100,000 were living in the more westerly territory acquired by purchase.

From the southern seaboard, the plantation system was being extended westward by planters in search of virgin land for cotton, the new staple of southern agriculture; and a mix of grain and livestock was becoming established as the agricultural base of the trans-Appalachian north. Products of the interior moved to the Gulf coast by way of the Mississippi River system, which had been brought under American control with the purchase of the Louisiana Territory, en route to the Atlantic seaboard and overseas markets.

As the trans-Appalachian resource region was opened to settlement, what Lampard (1955: 119) has called "colonialism on a continental scale" began to appear. The course of development in the interior was to be shaped in the port communities on the settled seaboard which a century earlier had been outposts in colonial resource regions organized through European metropolitan centers.

Chapter 3

CONTINENTAL COLONIALISM

If a local concentration of "full-time" traders distinguished the town from the village, the metropolis was distinguished by its cluster of business groups engaged in organizing and financing trade. The fortunes of the metropolis rested on the ability of these groups to interpose themselves between buyers and sellers, distant as well as proximate, in the money and commodities markets.

From an American perspective, one salient factor in the rise of European metropolitan centers had been the success of European-headquartered business groups in extracting a marketable surplus from what was previously unsettled and unexploited territory on the North American mainland. Another factor had been the success of these businessmen in directing trade flows generated within that territory through their home communities. Both elements were incorporated in strategies for creating a metropolitan organization centered within the United States of America.

American town life at the founding of the Union was concentrated around Atlantic ports which had been active in the colonial coastal and overseas trade. That part of the mainland integrated through the port communities into a European-centered metropolitan organization had extended inland only a few score miles from the Atlantic coast. To the west of the seaboard stretched a vast, essentially untapped territory. Development rights in this territory were being contested by business groups in the leading American port communities, as well as by foreign competitors.

A primary requisite of a metropolis for the interior was access to both the interior and the settled areas on both sides of the Atlantic in which the interior surplus could be marketed. Although the American ports were on the same continent as the interior resource region, they were more isolated from it than from overseas areas bordering the Atlantic. Water transportation was the only practicable means of large-scale, long-distance movement, especially for the bulk commodities of low per-unit value which were likely to constitute the initial exportable surplus of the interior. There was no direct water route between the interior and the Atlantic seaboard.

At the outset, the competitive positions of American-headquartered business groups with respect to interior development were defined primarily by the locations of their respective home communities with respect to a configuration of geographic features. Internal transport improvements soon modified the access advantage of one port relative to another, but the improvements were themselves constrained by the configuration, which is best described graphically (Figure 3-1).

The location of the Appalachian Mountains is outlined by the configuration of heads of rivers draining to the Atlantic and the Gulf of Mexico, respectively.

Within the transmontane region, the major water route was the Ohio River and its tributaries which flowed into the Mississippi River and ultimately drained into the Gulf of Mexico. The northern limits of the region were the shores of the Great Lakes.

Arrayed from north to south along the Atlantic coast were the ports of Boston, New York, Philadelphia, Baltimore, and Charleston. Around them were the largest nucleated settlements in the new nation.

The northernmost break in the mountain barrier was the Mohawk Valley, due west of Boston. Although the break was equidistant from Boston and New York, the access advantage rested with New York which alone had a water route to the valley.

A few hundred miles to the south and due west of Baltimore was another break in the mountain barrier, the Cumberland Gap. Most centrally located with respect to the Gap, and the headwaters

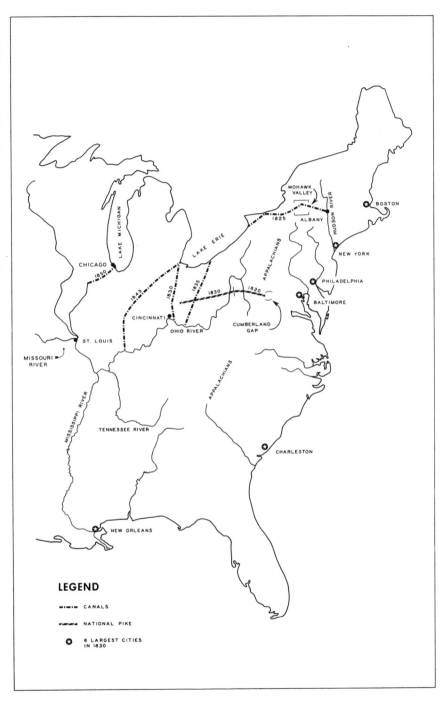

Figure 3-1. Selected features in city locations under a water-ways regime: United States to 1850.

of the Ohio as well, were Philadelphia and Baltimore, with the overland distance giving Baltimore a slight advantage.

The location of Charleston was eccentric with respect to the main thrust of interior development along the Ohio River. Moreover, the plain which lay to the west of the port was cut by rivers which funneled traffic to other coastal outlets.

Internal transport improvements were only beginning to shift the competitive positions of the Atlantic ports relative to one another when, as a group, they faced a new rival in organizing and financing interior trade. With the Louisiana Purchase, the United States acquired another established seaport community, New Orleans. Even before the Gulf port at the mouth of the Mississippi River came under American control, the flow of products downstream from the interior had been sufficient to make New Orleans prices newsworthy in the trans-Appalachian region. Now steamboats were overcoming the problem of upstream movement (see Wade, 1959: 26, 190).

AN AMERICAN METROPOLIS

Internal transport improvements first shifted the trans-Appalachian access advantage to Baltimore. On grounds that an east-west route was in the national interest militarily and economically, the federal government authorized construction of an improved road leading westward from the Cumberland Gap. The road was to be financed from proceeds of the sale of federal land beyond the Appalachians. This "National Pike" had reached the Ohio River by 1820 and become a major artery of internal movement.

The access advantage of Baltimore was short-lived, for it was through the Mohawk Valley that a direct water route between the Atlantic seaboard and the transmontane region first was completed. By 1825 products could move from the Great Lakes to the port of New York by way of the Erie Canal and Hudson River. Within a decade, canals had been cut which linked the Ohio River and Lake Erie.

New York scarcely had consolidated its position as the Atlantic focus of trade with the interior when railroads and steam locomotives appeared on the American scene. Rival Atlantic ports began to extend rail lines toward breaks in the mountain barrier, but before these lines penetrated the interior, rails were in place along the route of the Erie Canal. Shortly after 1850 rail connections were in place between Philadelphia and Baltimore, respectively, and the Ohio River and between Charleston and the Tennessee River. A northerly rail route by then linked New York with the tip of Lake Michigan, however; and the access advantage remained with New York as the waterways regime drew to a close.

New York's role as metropolis with respect to the interior cannot be accounted for solely in terms of a trans-Appalachian access advantage over other Atlantic ports. The Gulf port of New Orleans at the mouth of the Mississippi River offered an alternative outlet for interior trade and, indeed, became second only to the port of New York in shipping activity. In fact, Vernon (1963: 30) has observed that "Historians are a trifle unsure why New York lunged forward as the nation's first city in the early part of the nineteenth century" although its lead in foreign trade can be documented: a tenth of the national total in 1800; two-fifths by 1830; and three-fifths by the close of the waterways regime (Vernon, 1963: 31–32).

New York's unique advantage may have been a combination of: high accessibility to the national market of consumers of imports; control of a water route for movement of an exportable surplus from the interior; and a nucleus of mercantile institutions and an accumulation of mercantile capital dating from the colonial period. Philadelphia and Baltimore rivaled New York with respect to accessibility to the consumer market, and Philadelphia and Boston shared with New York a well-established mercantile core. New Orleans was the sole rival of New York with respect to water access to the interior. The coincidence of advantageous factors distinguished New York.

It is not clear that any other of the ports was even a serious contender as a "metropolis" for the interior. There was no substantial trade flow between the trans-Appalachian area and either Boston or Charleston. The gross flow of trade between Philadelphia, Baltimore, or New Orleans and the interior was more substantial, but trade relations were asymmetric. Imported or indigenous manufactures moved overland from Philadelphia and Baltimore to the interior, but overland transit costs prohibited movement of bulk materials from the interior to these mid-Atlantic ports. Bulk materials moved in substantial volume from the interior along waterways leading to New Orleans, but the eccentricity of the port with respect to both the supply area and the consumer market for manufactures damped the movement of finished goods into the port.

Business groups at the port of New York, the nation's foremost center of foreign trade, played a major role in setting the terms under which the exportable surplus of interior resource regions was marketed, the requisite imports of the interior were obtained, and interior producers could secure credit. New York's role with respect to the interior was that of the "classic" metropolis, the role played by London with respect to American resource regions on the Atlantic seaboard some decades earlier.

An imagery of interior producers shipping raw materials to New York traders in exchange for finished goods captures the notion of mutual dependency between resource region and metropolis, for from the exchange derive the profits of both producers and traders. It is too simplistic as a description of the interlocking and often indirect dependencies among interior resource regions and the metropolis of New York, however. Enumeration of a few features of the interior economies can convey something of the complexity of the metropolis-region relation.

The most valuable American export was raw cotton, the primary output of the plantation system that stretched across the south from the Atlantic coast westward beyond the Mississippi (Figure 3-2). Its selling price depended on demand in an English-dominated market. The price of foodstuffs imported into the plantation districts from the trans-Appalachian north depended not only on extraregional domestic demand, but also on foreign demand. In fact, foodstuffs were second in value only to cotton in the American export profile. Demand in the thinly populated interior was only a minor factor in the pricing of finished goods imported from supply areas bordering both sides of the Atlantic, however. Hence, the profitability of interior economic activity was influenced heavily by the terms of a series of exchanges agreed upon by businessmen in New York and their counterparts in foreign metropolitan centers. The prosperity of the New York businessmen was linked, in turn, to the volume of trade generated within the interior and directed overseas.

To increase and sometimes even to maintain output, interior producers had to extend their holdings of land and slaves or stock.

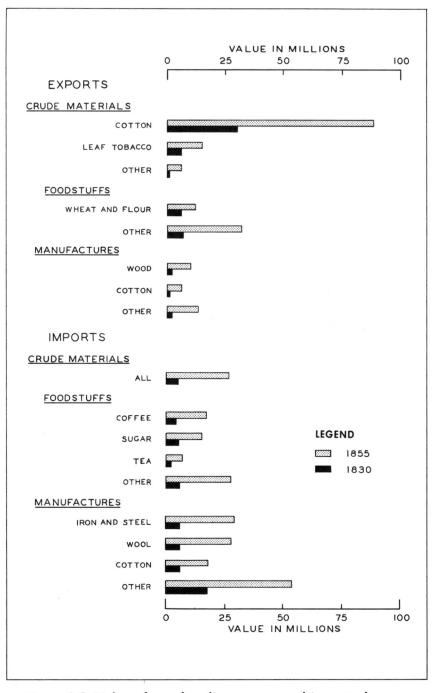

Figure 3-2. Value of merchandise exports and imports, by economic classes, for the United States: 1830 and 1855. (Source: *Historical Statistics of the United States,* Series U61–115.)

Their existing holdings represented the past accumulation of capital in the interior; extension required profits from current sales and credits against future sales. Again, the fortunes of the resource region and the metropolis were interdependent.

Organization of the interior resource region was the distinctively metropolitan function of New York, but it was not the sole commercial function. The city was also the primary trade center for the surrounding district. Its own population constituted the nation's largest concentrated market: 100,000 in 1810; 250,000 in 1835; 500,000 by 1850. Commodities of local and distant origin were brought together in the city for dispersal among buyers in the environs of New York. Locally produced commodities were collected for exchange within the district or shipment to other trade centers. Moreover, New York was a transfer or mixing point for traffic between other cities. Sea lanes, internal waterways, roadways, and rail lines converged on or radiated from the New York district. Such a coincidence of commercial functions—local market place, transshipment point, and coordinating center for long-distance trade— was, in fact, the hallmark of the metropolis as it had evolved in Western Europe.

Although New York alone qualified as a "national metropolis," other outstanding commercial centers were developing on the American continent, both along the coast and inland. Trade generated within a number of subnational territories had become sufficient to support regional centers which might qualify as metropolises if a finer-grained grid were imposed on the spatial structure of the national economy. Boston was local market place, transshipment point, and coordinating center for long-distance trade within New England; New Orleans performed these functions for the lower Mississippi Valley. A "regional metropolis," however, played a significant role in the economic development of only a section of the nation and, even within this limited territory, shared control of economic activity with the national metropolis of New York.

RAILS REINFORCE WATERWAYS

The port function and ancillary mercantile services had been the keystone of the classic metropolis. Railways in no sense diminished the strategic position of the seaports with respect to overseas trade and, in fact, allowed them to perform traditional "port" services for inland territories to which they lacked direct water access. Early rail building took the form of constructing short segments of road along which a steady flow of traffic could be anticipated, each segment separately financed and operated. Philadelphia, for example, had connections with another seaport, Baltimore; with an internal waterway to its west, the Susquehanna River; and with the anthracite coal fields some hundred miles northwest. Private, state, and municipal sources were providing sufficient capital to build railroads from established trade centers to traffic-generating sites. The center-oriented commercial network which had developed under a waterways regime was being reinforced with rails.

As late as 1830, six seaports which had been active before the formation of the Union were the sites of the largest American communities. New York had established a commanding lead over second-ranking Philadelphia; Baltimore had displaced Boston from third to fourth rank; and New Orleans had shifted ahead of a static Charleston (Figure 3-3). Their ranks with respect to population size approximated an ordering by commercial importance, for commerce was the city-building industry.

The first cities to break into the ranks of the six established centers were neither rival seaports nor new centers created by the spread of the rail net. They were places which occupied strategic locations on internal waterways along which an increasing volume of traffic moved as the interior was organized. (Growth patterns are displayed in Figure 3-3. Locations with respect to the waterways are shown in Figure 3-1.)

At Albany had stood Fort Orange, planted by European colonizers at the head of navigation on the Hudson River. The waterway junction gained importance when New York businessmen secured a direct route to the interior, for from Albany led the Mohawk Valley break through the mountain barrier. By 1840 Albany outranked Charleston, the smallest and most slowly growing of the established centers. Albany itself was outranked not only by the larger seaports, but also by Cincinnati, the "boom town" of the interior, however.

Cincinnati was a true city of the interior, located midway down the Ohio River in the heart of the trans-Appalachian resource

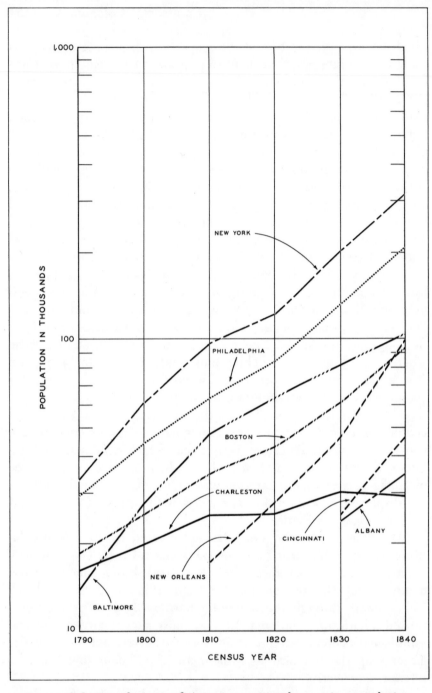

Figure 3-3. Populations of American cities larger in population
size than Charleston, S.C.: 1790 to 1840.

region. Its origins traced back only to 1788, and as late as 1820 its residents had numbered fewer than 10,000. Livestock and grain from the surrounding agricultural district were brought to local slaughterhouses and mills; agricultural commodities were shipped from the river port toward their ultimate markets. Goods brought from the east to the headwaters of the Ohio moved downstream to Cincinnati where they were mixed with commodities moved upstream from New Orleans and, then, dispersed through the surrounding territory. Cincinnati had, by 1840, become established as the primary collecting, processing, and distributing center in the interior and as the sixth-largest American city. (Development of the commercial nucleus in Cincinnati has been detailed by Wade, 1959.)

A record thousand miles of railroad were built within the nation in 1848. Steam locomotives had been "made in America" for more than a decade. The domestic manufacture of iron rails had begun. A multiplicity of ownerships and differences among operating companies in track gauge and equipment meant frequent tolls and transfer charges, substantial losses of time in handling. Nonetheless, demand for railroads was growing in all sections of the nation; and 1850 marked the first extensive grant of land from the public domain to further railroad building. Proceeds from the sale of land granted to the states of Illinois, Mississippi, and Alabama were to finance a rail route from Lake Michigan to the Gulf. The Gulf port of New Orleans was linked by rail with the tip of Lake Michigan in less than a decade. Still, however, rail building was concentrated in settled territory east of the Mississippi and oriented to centers of commerce which had become established under a waterways regime.

The roster of the top five American cities had remained unchanged from 1820 through 1860. New York's position as the nation's first city remained unchallenged, and a continuation of growth at the 1850s level would bring its population to the million mark before 1870. Philadelphia with more than a half-million residents was firmly positioned in second rank. Less than half the size of Philadelphia was third-ranking Baltimore, which held only a slight edge over fourth-ranking Boston and fifth-ranking New Orleans. Lagging New Orleans in population size by fewer than

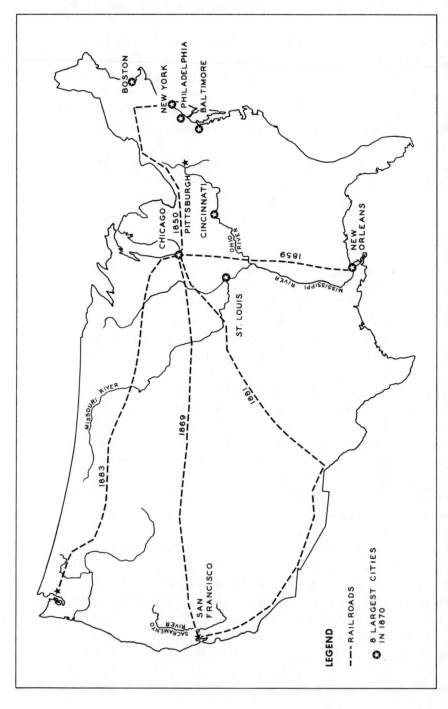

Figure 3-4. Selected features in city locations west of the Appalachians under a railroads regime: United States in the late 1800s.

LEGEND

--- RAIL ROADS

✪ 8 LARGEST CITIES IN 1870

10,000 residents in 1860 were two cities of the interior, Cincinnati and St. Louis, however.

St. Louis was an even "newer" city than Cincinnati and stood further to the west on the edge of the trans-Mississippi territory. Although settlement at the site had been continuous since 1764, not until the 1830s did the community have as many as 10,000 residents. Reflected in its rapid growth since that time were the opening of trans-Mississippi resource regions and the extension of the national territory to the Pacific coast. Under the waterways regime still effective beyond the Mississippi in 1860, St. Louis was the gateway from the settled east to the sparsely populated and largely unexploited west. The access route began just above the city at the point where the Missouri River entered the Mississippi, proceeded westward along the Missouri for a straight-line distance of some 250 miles, and continued either northwesterly along the course of the river or to the west and southwest by overland trails such as the Santa Fe. (The position of St. Louis in relation to the Mississippi River system is shown in Figure 3-1.)

Demand was growing for rail lines which would span the trans-Mississippi territory and link the newly acquired territory west of the continental divide with the settled east. Only limited capital was available from private and local government sources for construction through territory with an untested capacity for generating traffic. Federal assistance for so speculative a venture was needed, but action was stalled by the conflict of sectional interests with respect to the route that a transcontinental line should follow.

With the secession of southern states from the Union, federal support for rail building in the west came quickly. Military and political considerations coincided with economic considerations favoring a connection between the northeast and the growing Pacific-coast settlement centered on San Francisco Bay. In 1862 the Congress made a direct grant of land from the public domain to companies which would undertake to build a railroad between the Missouri River in Iowa and the Sacramento River in California. Seven years later the route was open (Figure 3-4).

New York and Philadelphia retained their positions as the nation's first and second cities, respectively, as late as 1880; but Baltimore, Boston, and New Orleans failed to maintain their lead over rival centers. The modest growth recorded in Cincinnati was

sufficient to shift it ahead of the Gulf port of New Orleans. A more
substantial gain shifted St. Louis ahead of Baltimore, the southern-
most of the major Atlantic ports, as well as New Orleans. Chicago
shifted ahead of New Orleans, Cincinnati, and Baltimore in the
1860s and, then, displaced St. Louis and Boston in the 1870s to
become America's third-ranking city of 1880 (see Figure 3-5).

Chicago had been transformed from a village of 4,000 persons
into a city with more than 100,000 residents between 1840 and
1860. During these years, water routes from the community on the
southern tip of Lake Michigan to New York on the Atlantic coast
and New Orleans on the Gulf coast had been reinforced with railroads (Figures 3-1 and 3-4). The continuous settled area had
reached the western limit of the trans-Appalachian north, near
which Chicago lay, and was spreading across the Mississippi into
the west-central plain. It was then that the southern states seceded
from the Union, orienting the western territory to Chicago and
through Chicago to New York more abruptly and definitively than
might otherwise have been the case. Eastward and southward trade
flows directed to St. Louis and New Orleans were disrupted at the
same time that work began on a transcontinental railroad feeding
into the network centered on Chicago (Figure 3-4).

By 1880, two additional communities had penetrated the ranks
of the established centers; both San Francisco and Pittsburgh
shifted ahead of New Orleans in rank with respect to population
size during the 1870s (Figure 3-5). The rise of San Francisco into
the ranks of the established centers of commerce can be accounted
for within a framework of continental colonialism or the classic
relation between metropolis and resource region. The port had been
the gateway to the far west, where resource extraction was gaining
momentum, under a waterways regime; and its access advantage
over other Pacific-coast communities had been strengthened by the
positioning of the first transcontinental rail route (Figure 3-4). The
framework fails to accommodate the rise of Pittsburgh, however.
Situated at the headwaters of the Ohio River, on the eastern edge
of the trans-Appalachian north, Pittsburgh had established itself as
a trade and transshipment center; but it was a commercial center
distinctly subordinate not only to the Atlantic ports east of the
mountain barrier, but also to Cincinnati downstream on the Ohio
and Chicago, St. Louis, and New Orleans still further to the west

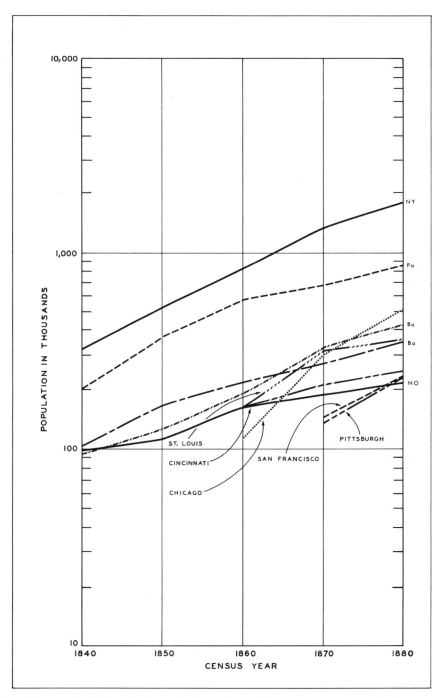

Figure 3-5. Populations of American cities larger in population size than New Orleans: 1840 to 1880.

(Figure 3-4). The popular designation of Pittsburgh as the "Iron City" suggests that heavy manufactures had become a city-building industry, a supplement if not alternative to the commerce generated by resource extraction.

Although continental colonialism cannot serve as an exclusive framework in accounting for the rise of centers to national prominence after the 1870s, it is not necessarily an unserviceable framework thereafter. Decennial gains in population in the trans-Mississippi plain were only beginning to match those recorded in the interior east of the Mississippi or along the Atlantic seaboard; the territory west of the continental divide remained largely unsettled. Extraction of known resources in the west was not yet being pursued intensively, and the resource potential had scarcely been explored. No major commercial centers were to be found west of the Mississippi Valley, save San Francisco on which an isolated seaboard settlement was centered.

During the next four decades, eight new centers were to penetrate the ranks of the established centers. Five of the new centers were located in long-settled territory east of the Mississippi and were encircled by established centers of commerce. Minneapolis, Kansas City, and Los Angeles, however, were western cities; and in accounting for their rise, the classic metropolitan formulation remains potentially relevant.

Chapter 4

STEAM, STEEL & METROPOLITAN REORGANIZATION

The beginnings of the American factory system can be traced to the manufacture of cotton cloth in New England in the 1810s. Lampard (1955: 118) has argued, "The impetus toward the 'integrated' factory . . . came not from the semi-rural manufactures but from the embattled commercial metropolis" of Boston. While mercantile capital in New York, Philadelphia, and Baltimore was being diverted into building routes to the interior and organizing agricultural specialties beyond the Appalachians, Boston businessmen—isolated from the main line of westward development—were financing manufacturing plants in New England.

Steam and steel had no key role in the early phases of industrialization in the manufacturing sector although by the 1870s they had become prominent features of the American industrial scene. The use of steam power in the manufacturing process or an areal concentration of steel making is perhaps a sufficient, but certainly not a necessary indicator of the onset of industrialization. Shimkin (1952: 89) has identified the generic features of industrialization as: progressive systematization and simplification of work processes; extensive division of labor; substitution of mechanical for human effort; and the development of large controllable sources of supply. Prerequisite are "mechanisms promoting communication, training, experimentation, and habit change" and "an organization permitting wide exchange, effective cooperation," as well as "appropriate physical resources." From the steam-and-steel complex, nonetheless, came the fundamental change in the territorial basis of American economic activity which in common parlance is said to be the outcome of the industrialization of manufactures.

Reorganization in the intercommunity division of labor between countryside, town, and metropolis did not follow directly from the introduction of steam locomotives and steel rails as an alternative transport mode. The relative numbers of resource extractors, traders, and financiers continued to differentiate the community types. The rail system was becoming most efficient, however, as a long-haul, heavy-load transporter; and extension of the system was generating an unprecedented demand for products of the iron and steel industry. These factors combined with the operating characteristics of the steam engine as a power source in the manufacturing process to create large-scale concentrations of heavy manufactures. The loss of efficiency in power transmission over a distance by a shaft-pulley-belt system called for a clustering of machines near the engine, and the per-unit power costs varied inversely with the scale of production. The substantial material inputs and sizable market required to support so large a manufacturing enterprise could be reached through the spreading rail network. The manufacturing concentrations might appear spatially distinct from or coincident with the established metropolitan centers; in either case, a new kind of city was created.

At this point in the American history, a "system" of great cities differentiated by functional type as well as scale of activity first can be discerned. Few if any members performed the full range of non-extractive or urban functions. Each member was industrially specialized and depended upon other members of the system to perform essential complementary functions. Some established metropolises added large-scale manufactures to their economic bases; others did not. In some new centers, the traditional metropolitan pursuits loomed large in the industrial profile; the profiles of other new centers were dominated by large-scale manufactures. The intercommunity division of labor now was being extended into the top ranks of the urban hierarchy.

THE SETTING

The stretch of Atlantic seaboard bounded by Boston and Baltimore was a "natural" location for most small-shop and factory manufactures that had developed by 1860. Accessibility to mercantile capital and entrepreneurial experience, to manpower and the market, was higher here than in other sections of the nation; and material inputs could be assembled here as readily as elsewhere. Concentrated in New England and six neighboring mid-Atlantic states was 70 percent of the national manufacturing work force, in which workers in the boot-and-shoe, cotton-goods, and men's-clothing branches were most numerous (Perloff et al., 1960: 119–120). The factory system had appeared earlier and become more pervasive in the cotton-goods branch of manufacture than in the boot-and-shoe and men's-clothing branches. In the latter branches, the bulk of the work continued to be carried out in small shops although specialties such as cutting had been centralized, stitching machines were in use, and the "ready-made" market was growing by 1860. Workers engaged in each branch numbered nearly 125,000, but their workplaces were too widely dispersed to give rise to great cities based on the manufacture of textiles, footwear, or apparel. Even the concentrations of clothing workers around the seaport centers were little more than the traditional clustering of finished-goods producers in the environs of a metropolis.

Interspersed among the men's-clothing, cotton-goods, and boot-and-shoe branches of manufacture in the top five ranks with respect to value added were two processing industries, the lumber and the flour-and-meal industries (Perloff et al., 1960: 119). Unlike the apparel, textile, and footwear branches, the lumber and grain-milling industries had as inputs primary raw material which underwent a substantial weight loss in processing. Such processing industries were attracted to sites near the sources of their raw material inputs, especially those sites along transport routes to major markets. The "pull" of transmontane supply areas was, by 1860, reflected in the westerly distribution of lumber production and grain milling vis-á-vis the manufacture of apparel, textiles, and footwear.

The demand for wood both as fuel and fabricating material had nearly depleted the forest resources of the northern Atlantic seaboard. It was now being met by lumbering in the trans-Appalachian north where forest resources were abundant and transport routes to the seaboard were direct (Perloff et al., 1960: 215). Grain milling no longer was concentrated in seaboard centers such as Philadelphia and Baltimore, but was to be found as far west as St. Louis. Wheat growing had spread from the mid-Atlantic seaboard westward to the trans-Mississippi plain. Added to the market area which lay along the Atlantic was the plantation district extending through the south-central section of the nation (Alderfer and Michl, 1957: 501).

"Not until the decade between 1860 and 1870 did it become apparent that the complete supply of staple products for the home market was within the capacity of domestic manufactures. During the Civil War, the great demand for manufactured supplies of every description and the high protective duties on imports necessitated by the revenue requirements of the Government stimulated enterprise and production to an extent not known before or since" wrote a statistician of the Bureau of the Census in 1902 (Twelfth Census, Vol. 7, Pt. 1: 1). Statistics on the value of or capital in domestic manufactures at this time of rapid expansion are difficult to interpret, for the United States abandoned specie backing for its money from 1862 to 1879. Nonetheless, reports of the Census of 1880 make it clear that the iron and steel, and the foundry-and-machine-shop-products branches of manufacture had gained prominence in the national industrial profile during these years.

In 1870 activity in the iron and steel industry was concentrated in the neighboring states of New York, New Jersey, Pennsylvania, and Ohio. New York and New Jersey soon were to be displaced from their initial ranks of third and fourth among the states with respect to tonnage produced, but Pennsylvania and Ohio were to remain the first- and second-largest producers, respectively, for some decades. Gaining ground rapidly as a producer during the 1870s was the state of Illinois, which by 1880 had established itself in fourth rank. The location of supply and market areas and the placement of transport and communication networks combined to the advantage of transmontane territory north of the Ohio and east of the Mississippi as the site of production.

Supply areas alone did not define the production sites, for "all of the raw materials essential to a full development of the different branches of the industry" were not available in the same locality.

In the southern states were "vast deposits of iron ore, coal, lime-stone, and dolomite, all lying within short distances of each other"; but these ores were "not adapted to the manufacture of pig iron suitable for making steel by the Bessemer process, by which in the census year 1900 almost three-fourths of the crude steel made in the United States was produced." Michigan, Wisconsin, and Minnesota contained "immense deposits of iron ore," much of which was "especially adapted for the manufacture of pig iron suitable for making Bessemer steel"; but none had "a mineral fuel supply that could be economically used in the manufacture of pig iron." Pennsylvania was "endowed by nature with boundless stores of both anthracite and bituminous coal"; but the state was "compelled to draw largely upon other sections of the country for the ore for its blast furnaces, the greater part coming from the Lake Superior region over 1,000 miles away." Although Illinois was "located much nearer than Pennsylvania to the Lake Superior iron-ore mines," the state had to obtain "the bituminous fuel for its blast furnaces from the Connellsville region in Pennsylvania, and from the Pocahontas Flat Top region in West Virginia, hundreds of miles away." New York and New Jersey used Lake Superior ore and Connellsville coke, "frequently mixing the coke with anthracite coal for blast furnace use." (All quotations from Twelfth Census, Vol. 10, Pt. 4: 20–21.) The supply area for the industry extended even beyond the national territory. In 1900, for example, foreign ore made up 3 percent of the total ore consumed by tonnage and 6 percent by cost. Half the imported ore was obtained from Cuba, and most was consumed by blast furnaces in Pennsylvania and states bordering the Atlantic (Twelfth Census, Vol. 10, Pt. 4: 31).

The extreme localization of production in specific branches of the iron and steel industry makes it clear that for many establishments the market was near national in scope. The state of Pennsylvania, for example, produced 49 percent of the rails in 1880 and 57 percent of the rails in 1900 although extensive railroad construction was not then occurring within the State. Pennsylvania establishments in 1900 produced: 81 percent of the car axles; 86 percent of the Bessemer steel shapes; 93 percent of the open-hearth steel shapes; and all the steel armor plate and gun forgings. (The examples are drawn from Twelfth Census, Vol. 10, Pt. 4: 59 ff.) The consumers of products of the iron-and-steel industry included few households or individuals, but the distribution of population in the late 1800s can bound the territory within which fabricators or other industrial consumers would choose to locate. Distributions of the mileage of railroads constructed and operated can define territorially one segment of the market, as well as depict the placement of a major component of the internal transport network.

The most populous section of the nation in 1860 had been the territory on the northern seaboard, and increases in the population of the north Atlantic section over the next four decades matched, in absolute terms, the increases recorded in any other section of the nation (Figure 4-1). The area north of the Ohio between the Pennsylvania coal fields and the Lake Superior ore reserves had become the second most populous section by 1860, and subsequent gains in the east-north-central section were as great as those recorded in the north Atlantic section. Equally substantial gains recorded in the west-north-central plain shifted that section from fifth to third position in rank with respect to population size between 1860 and 1900, displacing in rank both the south Atlantic and east-south-central sections.

Most of the railroad track laid before 1870 was east of the Mississippi, and the network was far denser in the north than in the south. All sections of the nation shared in the rail-building boom that peaked in the 1880s, but the mileage added was greater in the plains and the transmontane north than elsewhere. By 1890 the railroad mileage in operation was as great in the plains as in the transmontane north and exceeded the operating mileage in either the northeast or the southeast where rail building had first occurred. Filling in the skeleton network spanning the southeast accounted for more railroad construction after 1890 than did the extension of lines into unpenetrated western territory, and when rail building slackened after the turn of the century, the operating mileage still was heavily concentrated in the eastern half of the nation. Pronounced sectional differences in the timing of railroad building are brought out in the statistics on mileage in operation reported by Perloff et al. (1960: 196) and summarized below (Table 4-1).

The carrying capacity of the railroads was increasing more rapidly than time series on mileage in operation suggest. Loads handled increased to as much as 4,000 tons, a twentyfold increase,

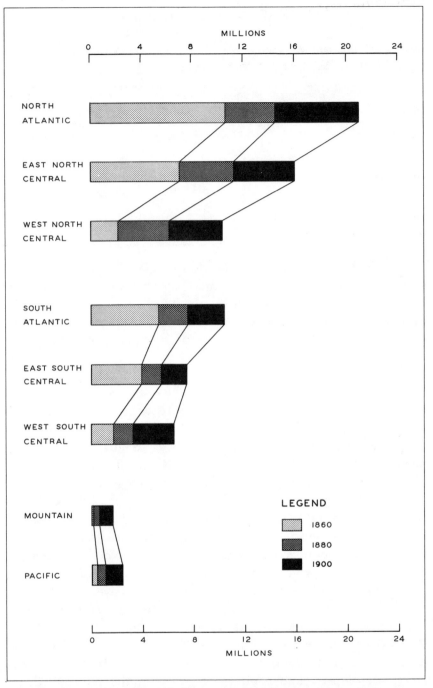

Figure 4-1. Population by geographic division: United States, 1860 to 1900.

Table 4-1. Railroad Mileage (in thousands) Operating in 1870 and Added
1870–1910, by Section of the United States

Section	Operating 1870	Added 1870–1890	Operating 1890	Added 1890–1910	Operating 1910
Northeast	15	11	26	6	32
Great Lakes	15	22	37	8	45
Southeast	12	19	31	24	55
Plains	8	29	37	12	49
West	3	15	18	15	33
Southwest	1	12	13	13	26

as 250-ton locomotives built after 1900 replaced the 25-ton locomotives of the 1850s and steel rails replaced the iron rails laid
before 1870 (Johnson and Van Metre, 1921: 39–43, 47, 59). Moreover, the disruption of long-distance traffic by frequent transfers
between railways lessened. Specialized "transportation" companies
were established to handle interrailway services at a time when no
single railway company controlled as much as a thousand miles of
line and the railway companies, themselves, had not yet developed
a system of car interchange. As early as 1840, firms engaged in the
express business and operating over the railroads were active on
the Boston-New York route. By the 1850s freight dispatch companies were soliciting freight, providing cars for its shipment, offering special terminal facilities, and handling payments to the railways for haulage over additional routes. Consolidation of short
connecting rail lines under a single ownership also was easing
the frictions of transfer although it was not until the 1890s that
systems with control over 10,000 or more miles of line emerged
(see Johnson and Van Metre, 1921: 90 ff., 171–175, 203–205, 275).

Insofar as rapid telecommunication facilities were relevant in
the initial siting of large-scale manufacturing enterprises, the
medium of interest is telegraphy; and telegraph service had followed the spread of the rail network. The "dominant carrier in the
domestic telegraph industry," the Western Union Telegraph Company, "developed close contractual ties with the railways. Telegraph pole lines were constructed along railroad rights-of-way. The

lines were used jointly for general telegraph and railroad telegraph communication and signalling; and railroad stations and personnel were used for the pick-up and delivery of telegraph messages" (*Historical Statistics of the United States,* 1960: 474). Although by 1900 over a million telephone instruments were in use in the nation and nearly 200,000 toll calls were made daily, telephony had developed too late to influence the spatial configuration of the first manufacturing centers (*Historical Statistics of the United States,* Series R1, R7, and R9).

By 1880 the iron and steel, and foundry-and-machine-shop-products industries would have been identified as "leading" branches of manufacture not only in the transmontane north, but in the American industrial structure as a whole. In 1900 they still would have been so identified. Under at least one system of industrial classification, each would have ranked among the top five branches of manufacture whether the criterion of ranking was number of workers, value added, or capital invested (Table 4-2). (Ambiguity in interindustry "size" comparisons is always present, for size de-

Table 4-2. Leading Branches of Manufacture in the United States, 1900

Industry	WORKERS, thousands 1880	1900	VALUE ADDED, millions 1880	1900	CAPITAL, millions 1880	1900
Boots and shoes, factory product	111	151	64	91	43	102
Cotton goods	185	308	97	163	220	467
Men's clothing	161	205	78	218	80	173
Lumber and timber products	148	296	87	249	181	611
Flouring and grist mill products	58	43	64	85	177	219
Foundry and machine shop products	145	378	111	359	155	665
Iron and steel	141	232	105	282	231	573
Printing and publishing	59	201	58	260	63	293
Malt liquors	26	47	44	186	91	415
Slaughtering and meat packing	27	80	36	103	49	191

Note: Value added calculated as value of products less cost of materials used.
Source: Twelfth Census, Vol. 7, Pt. 1, General Table 1.

pends upon the aggregation procedures followed. The boot-and-shoe industry, for example, is represented in Table 4-2 by the factory-product branch in which 151,000 workers were employed in 1900. An additional 35,000 persons were then working in five closely related branches, specifically, boot-and-shoe "cut stock," "findings," "uppers," and boots-and-shoes "custom work and repairing," and "rubber." The time dimension is a further complication in that a shift among industries with respect to size may reflect nothing more than a differential in the timing of change in the manufacturing process such as, for example, a shift from small-shop to factory production.)

Industrialization had begun earlier and proceeded further by 1900 in the cotton-goods branch than in the other manufacturing industries which had been included in the top five in 1860. Evidence of industrialization had become nearly as pronounced in the foundries and machine shops, which had more recently become prominent features of the industrial scene, as in the cotton-goods factories. Evidence of industrialization was even more pronounced in the iron and steel branch in 1900. In fact, of the five branches engaging the largest work force in 1900, the iron and steel industry ranked first on: capital in machinery, tools, and implements per worker; horsepower per worker; percentage of aggregate horsepower obtained from steam engines; percentage of the work force employed in large establishments; the ratio of salaried officials and clerks to proprietors and firm members; and the percentage of value produced by incorporated companies (Table 4-3).

Were a single indicator of industrialization in a branch of manufactures to be relied on, it would perhaps be the distribution of the value produced by the form of organization of the producing establishments. Incorporated companies, as opposed to firms and limited partnerships and individuals, accounted for 94 percent of the value produced by the iron and steel industry in 1900. Moreover, 28 percent of the total value produced by the industry was contributed by forty industrial combinations which together controlled 447 plants, 106 of which employed more than 500 workers and thirty-five of which employed more than 1,000 workers. Shortly after the census year closed, the United States Steel Corporation was formed, bringing together eleven constituent companies of which nine had themselves been included in the count of indus-

Table 4-3. Selected Characteristics of Leading Branches of Manufacture in the United States, 1900

Characteristic	Cotton Goods	Men's Wear[a]	Lumber Products	Machine Products	Iron, Steel
Capital in machinery, tools, and implements per worker	597	38	380	403	771
Horsepower per worker	2.7	0.1	4.8	1.1	7.2
Percent of aggregate horsepower from:					
Water wheels	31.1	1.8	12.5	4.8	0.5
Steam engines	65.5	44.2	86.9	75.1	94.7
Estimated minimum percent of workers in establishments with more than 100 workers[b]	94.6	29.7	29.5	51.8	96.5
Percent of workers classified as:					
Proprietors and firm members	0.1	5.5	12.8	2.4	0.1
Salaried officials and clerks	1.6	7.2	3.6	7.1	3.9
Percent of value produced by establishments with given form of organization:					
Individual	5.0	18.8	27.0	11.2	0.8
Firm and limited partnership	5.1	62.1	25.3	15.0	5.6
Incorporated company	89.9	19.1	47.7	73.8	93.6

a. Factory product, only.
b. Estimated from distribution of establishments by average number of employees.
Source: Twelfth Census, Vol. 7, Pt. 1, General Tables 2, 8, and 11.

trial combinations in the iron and steel industry in the census year. (All data from Twelfth Census, Vol. 7, Pt. 1: lxxvii ff.)

Industrial manufactures were becoming a major sector of activity in the national economy, supplementing the commerce and resource extraction which had dominated the industrial structure as late as 1860. In no branch of manufacture had industrialization proceeded further than in the iron and steel industry. In no section of the nation was the impact of industrial manufactures on the local economy more pronounced than in the transmontane north. Realignment among the established metropolises and lesser centers was inevitable, for some were positioned far more favorably than others to share in the manufacture and marketing of iron and steel and their products.

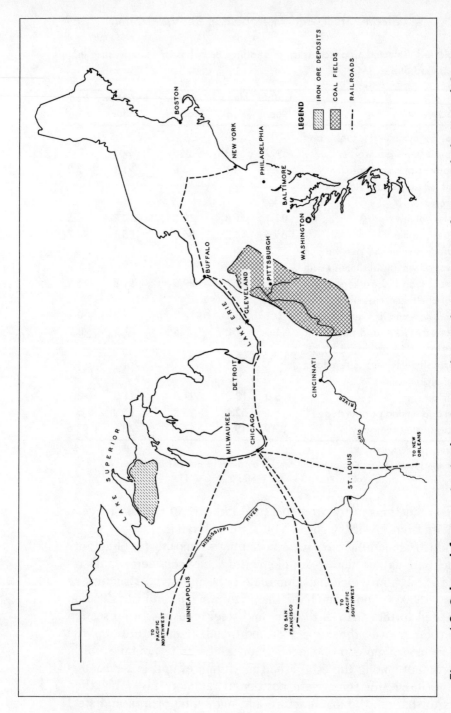

Figure 4-2. Selected features in city locations in the northeastern manufacturing belt: United States in the late 1800s.

ESTABLISHED METROPOLISES
AND NEW CENTERS

The entry of Pittsburgh into the ranks of the leading centers of population in the 1870s signals the beginning of metropolitan reorganization. The territory not only proximate to Pittsburgh but throughout the eastern half of the nation already had been integrated into the commercial network focused on New York, both directly and through the regional metropolises that encircled the city of Pittsburgh. Insofar as a swelling of commercial activity might account for Pittsburgh's favorable growth differential, the established metropolitan centers had lost "control" over either some territory once tributary to them or some economic activities carried on in the still tributary territory. In either case, the metropolis-region relation had been modified. If it were not a swelling of commerce, but rather the expansion of local manufactures that accounted for Pittsburgh's favorable growth differential, a new kind of city had emerged which matched the regional metropolis in size but differed from it in function.

Only a decade after Pittsburgh had entered the ranks of the leading centers, two cities in transmontane territory to the north of Pittsburgh on the Lake Erie shoreline—Buffalo and Cleveland—displaced an established regional metropolis in rank with respect to population size. In the following decade, two major centers emerged further to the west on the Great Lakes system—Detroit and Milwaukee. A chain of five new centers now stretched across the territory bounded on the east by the Pennsylvania coal fields and on the west by the iron-ore reserves near Lake Superior (Figure 4-2).

New Orleans and San Francisco were so remote from the transmontane north that metropolitan reorganization in that section could scarcely have influenced their roles as the regional metropolises of the south center and far west, respectively. Their isolation from the emerging "manufacturing belt," in which unusual numbers of manufacturers obtained their material inputs from and directed their outputs to other manufacturers, also precluded effective competition for "manufacturers' manufactures." Boston, Baltimore, and St. Louis lay closer to the manufacturing belt; but their locations were perhaps more eccentric with respect to the main

axis of development than were those of New York, Philadelphia, and
Cincinnati. Most favorably situated of the established metropolises
was Chicago (again, see Figure 4-2).

Timing and location suggest that the new transmontane centers
gained national prominence by capturing a disproportionate share
of the growth in the manufacturing sector, in particular, the metals
industries. As a quick independent check, the industrial structures
of the new centers can be compared with those of the established
metropolises on a single criterion: the percentage of gainfully
occupied residents who were "iron and steel workers" as of 1900.
The relative numbers range from 2.4 to 7.4 percent in the five new
centers, from 0.4 to 1.8 percent in the nine established centers and
Minneapolis, the new rail gateway to the northwest.

A local commercial core was not absent in the new transmontane
centers. In fact, with four to six "bankers, brokers, and wholesalers"
per 1,000 gainfully occupied residents, Pittsburgh, Buffalo, Cleveland, Detroit and Milwaukee were as specialized in banking,
brokerage, and wholesaling as were Boston, Philadelphia, Baltimore, Chicago, and Cincinnati. These centers were less specialized,
however, than were New York, New Orleans, St. Louis, San Francisco, and Minneapolis, in which the relative numbers of bankers,
brokers, and wholesalers ranged from seven to thirteen. It is, then,
the metal-working rather than the commercial specialties that
sharply distinguish the new transmontane centers. (Measures on
local industrial structures are based on data of the Twelfth Census,
Vol. 2, Pt. 2: 550 ff.)

(A comparison between centers based on the percentage of
gainfully occupied persons in each center who are, say, bankers is
not to be confused with an intercenter comparison based on the
percentage of the nation's bankers who are located in each center.
In the first instance, centers are contrasted with respect to the importance of banking in the local economy, "specialization" in banking. The latter contrast is in terms of the importance of the respective centers in the national banking system, "localization" of
banking. Chicago and Pittsburgh, for example, have been said to be
similar in their specialization in banking, brokerage, and wholesaling. Because Chicago's gainfully occupied residents and, hence,
bankers, brokers, and wholesalers outnumber those in Pittsburgh by
a factor of three, however, Chicago was the more important center

of banking, brokerage, and wholesaling from a national perspective. The share of the nation's activity in these lines of commerce localized in Chicago was three times that in Pittsburgh. In contrasting new with established centers, equally strong specialization typically implies a lesser localization in the new center or equally strong localization typically implies a lesser specialization in the established center; for, on the average, residents will be less numerous in the new center than in the established center.)

As indexed by the number of industrial combinations headquartered locally, New York was the primary control center for most branches of industrial manufactures in 1900 (Table 4-4). Seventeen of the forty iron-and-steel combinations, twelve of the forty-nine food-and-liquors combinations, and forty-one of the

Table 4-4. Location of Headquarters of Industrial Combinations in Specified Industries, 1900

Industry and Number of Plants Controlled	MAJOR CENTER										Other City*
	Bo	NY	Pa	Ba	NO	SF	SL	Cg	Pi	Cl	
All Industries											
All	4	70	5	4	2	5	4	18	16	6	51
5 or fewer	1	25	4	1	2	1	0	6	4	2	25
6 or more	3	45	1	3	0	4	4	12	12	4	26
Iron and Steel											
All	1	17	0	1	0	0	0	3	4	2	12
5 or fewer	1	4	0	1	0	0	0	0	0	1	7
6 or more	0	13	0	0	0	0	0	3	4	1	5
Food and Liquors											
All	0	12	3	1	2	4	1	9	1	1	15
5 or fewer	0	5	3	0	2	0	0	4	0	0	10
6 or more	0	7	0	1	0	4	1	5	1	1	5
All Other											
All	3	41	2	2	0	1	3	6	11	3	24
5 or fewer	0	16	1	0	0	1	0	2	4	1	8
6 or more	3	25	1	2	0	0	3	4	7	2	16

* No headquarters reported in Cincinnati, Detroit, Milwaukee, or Washington; one only in Buffalo and Minneapolis, respectively. All others in cities smaller than New Orleans.

Source: Twelfth Census of the United States, Vol. 7, Pt. 1, p. 1xxxvi ff.

ninety-six combinations in other branches of manufacture reported
New York as the site of the head office. The number of plants
controlled from a single head office is not altogether satisfactory as
an indicator of scale of combination operations or the territorial
dispersion of the combination's operating units. Nonetheless, from
the fact that forty-five of the seventy combinations based in New
York controlled at least six plants and one iron and steel combination alone controlled sixty-five plants, it can be inferred that New
York businessmen were organizing manufacturing activity beyond
the immediate environs of the city. New York's role as the national
metropolis had not been weakened by the growth of industrial
manufactures and emergence of a manufacturing belt in the interior; for the "first city" had captured a substantial share of the
distinctively metropolitan activity generated by industrial manufactures.

With the exception of Chicago, the established regional metropolises were infrequently chosen as head-office sites by industrial
combinations. Philadelphia, which in the 1870s had been the second
largest American city, had but five combinations headquartered
locally in 1900; nor had any other regional metropolis save Chicago
more than five. Chicago, however, was the choice of eighteen industrial combinations, including three whose primary product was
iron and steel, nine in the food-and-liquors industry, and six engaged in other branches of manufacture. Twelve of the eighteen
were combinations controlling at least six plants, and a Chicago-
based food combination reported ninety-five operating units.

New York and Chicago were the only metropolises to capture
a sizable share of the coordinating activity required by industrial
manufactures and symbolized by the organizationally complex industrial combination. The other noteworthy localization of this
distinctively metropolitan function was in Pittsburgh, the oldest
of the new manufacturing-belt centers. Pittsburgh-based combinations numbered sixteen, only two fewer than the number based in
Chicago. Four iron and steel combinations, a food-and-liquors combination, and eleven combinations engaged in other branches of
manufacture had chosen Pittsburgh as the site of the head-office;
twelve of the sixteen were combinations controlling at least six
plants.

These few pieces of evidence suffice to illustrate the functional

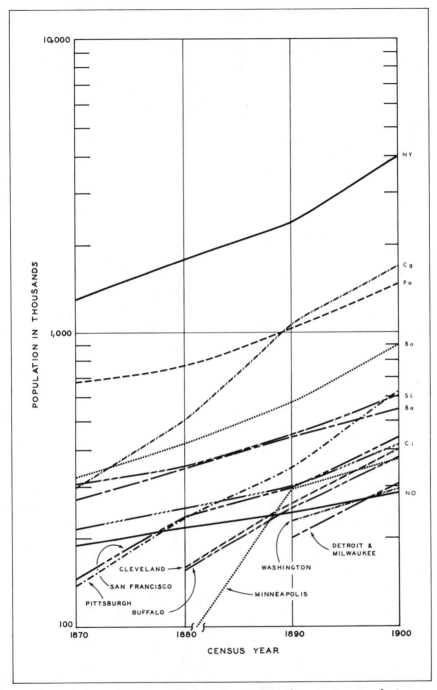

Figure 4-3. Populations of American cities larger in population size than New Orleans: 1870 to 1900.

differentiation among major population centers that followed from
their differential success in attracting producing or coordinating
activities associated with industrial manufactures. Traditional
metropolitan pursuits appeared still to dominate the industrial
profiles of some centers. In others, the profile was heavily weighted
with manufacturing activity. Still other centers were both sites of
large-scale manufacturing concentrations and focal points in extended commercial networks.

We will defer further description of the emerging system of cities
until we have examined more systematically the territorial organization of both industrial manufactures and the financial network at
the turn of the century. Before doing so, however, it is well to review the recent growth history of the major centers in the system
(Figure 4-3). The most noteworthy shift among the established
centers between 1870 and 1900 was Chicago's rise from fifth to
second rank with respect to population size. By 1900 the new manufacturing-belt center of Pittsburgh had established itself as the
nation's fifth-largest city, having displaced in rank five regional
metropolises since the 1870s. Still smaller in population size than
any established metropolis save New Orleans were the new manufacturing-belt centers of Buffalo, Cleveland, Detroit, and Milwaukee, as well as the new regional metropolis of Minneapolis. In
fifteenth rank was the national capital, Washington, which had
most recently displaced New Orleans.

The entry of Washington into the ranks of the leading centers
is, in one sense, a special case, in another sense, further evidence
of the finer territorial division of labor that was emerging. The
growth of paid civilian employment by the federal government in
the late 1800s seems adequate to account for the city's growth
(*Historical Statistics of the United States*, Series Y242). By 1900
there were twenty-two persons employed in Washington by the
federal government for every 100 gainfully-occupied persons living
in Washington. The increase in federal employees reflected, however, the increasing importance of regulation by the federal government in an economy where nationwide exchange of commodities
and information had become a key organizational element.

Chapter 5

THE MANUFACTURING BASE IN 1900

Natural resource extraction remained salient in the national industrial structure of 1900. Nearly two-fifths of the nation's workers were engaged in agriculture, forestry, fishing, or mining. Although a sixth of the nation's workers now were attached to the manufacturing sector, seven of every ten manufacturing workers were employed in an industry not more than two "stages" removed from natural resource extraction.

A sixth of the manufacturing workers, representing three per cent of the total work force, were employed in industries which might be identified as "first-stage resource users" because a relatively large share of their inputs were obtained from the resource extractors. The tobacco-manufactures industry was the leading employer in the group, followed by sawmills, planing mills, and millwork; meat products; and grain-mill products.

Just over half the manufacturing workers, representing nine per cent of the total work force, were employed in industries designated "second-stage resource users." These industries obtained a relatively large share of their inputs from the first-stage resource users producing for nonfinal markets, that is, for other industrial establishments which would further process or fabricate the products. Yarn, thread, and fabric mills; miscellaneous wood products; other primary iron and steel industries; and apparel and accessories were second-stage resource users. Other large employers in this resource-use group included: blast furnaces, steel works, and rolling and finishing mills; miscellaneous chemicals and allied products; furniture and fixtures; knitting mills; and bakery products.

Resource extraction had only indirect significance for the other manufacturing industries, which together employed 5 percent of the national work force; for a relatively small share of their inputs were obtained from resource extractors or first-stage resource users. Leading industries in terms of size of work force were: footwear, except rubber; railroad and miscellaneous transportation equipment; the miscellaneous manufacturing industries; newspaper publishing and printing; printing, publishing, and allied industries, except newspapers; and electrical machinery, equipment, and supplies. (The work force in non-electrical-machinery industries and, hence, this group as a whole is underestimated to some extent by our arbitrary assignment of the Twelfth-Census title "foundry and machine shop products" to the "other primary iron and steel industries," a second-stage resource user. Workers in "Foundry and machine shop products" establishments made up 1.3 percent of the total work force in 1900. Under current classification procedures, some of these workers would be assigned to the non-electrical-machinery industries.)

The classification of detailed manufacturing industries according to "stage of resource use" and "type of market" which we impose on the data collected in 1900 is identical to the classification employed by Duncan et al. (1960: 200 ff.) in the analyses of the 1950 industrial structure reported in *Metropolis and Region*. (Selected data from the *1947 Interindustry Relations Study* of the United States Bureau of Labor Statistics which are relevant to the classification appear in Perloff et al., 1960: appendix Table L.) The mid-century classification may fail to reflect faithfully an industry's resource-market relations in 1900, for change in the organization of the manufacturing process has occurred. It seems unlikely, however, that any substantial number of detailed manufacturing industries have been misclassified with respect to their relation to natural resource extraction or the type of market in which they disposed of their products at the turn of the century.

Selection of a resource-market classification for detailed manufacturing industries is actually a minor problem in organizing the 1900 data for analysis. Although our immediate objective is to compare the industrial structures of the major centers with one another and with the national structure as of 1900, we also are trying to establish a benchmark from which to trace subsequent change in the respective structures. There exists for 1900 no set of statistics on the industrial structures sufficiently comparable to contemporary data sets to permit the measurement of change. We believe, however, that by reorganizing data published in the Twelfth Census (1900), we have succeeded in establishing a fairly adequate baseline reading for the manufacturing sector.

Included among the publications of the Twelfth Census (1900) were two volumes which presented statistics for the nation and for places with populations of 20,000 or more on activity in specific branches of manufacture (Vol. 7, Pt. 1 and Vol. 8, Pt. 2). The branches of industry for which statistics were presented typically are sufficiently specific that each can be assigned a four-digit code of the current Standard Industrial Classification. Illustrative industry-branch titles might be: coffee and spice, roasting and grinding; liquors, malt; oil, cottonseed and cake; pottery, terra cotta, and fireclay products; sewing machines and attachments; and springs, steel, car and carriage. There is also, of course, a residual category of "all other" which cannot be assigned an industry code.

The first step in reorganizing the 1900 data was, then, to assign the appropriate current SIC (Standard Industrial Classification) code to each specific branch of industry. Assignment of industry branches to the detailed manufacturing industries used for tabulation in the 1960 Census of Population becomes straightforward, for the detailed industries are defined by the SIC codes of the component industries.

Employed workers in each detailed manufacturing industry are taken as the sum of "proprietors and firm members," "salaried officials, clerks, etc.," and "average number of wage earners," three items of information included in the Twelfth Census for each industry branch. The number of persons aged ten and over who were "engaged in gainful occupations" according to reports of the Twelfth Census are equated with the employed labor force in all lines of economic activity. Full comparability with the current industrial classification has not been effected, of course; but the match appears sufficiently close to provide a baseline from which change can be traced.

In retabulating the industry statistics from the Twelfth Census, data for the largest city in each major center of 1900 were combined with data for any cities with 1900 populations of 20,000 or more located in what now is identified as the "metropolitan ring" centered on the largest city. The composition of the 1900 center work force by place of residence for each of the seven centers including a sizable outlying community is:

Minneapolis—the city, 55 percent, and St. Paul, 45 percent;

Boston—the city, 58 percent, and Cambridge, Lynn, Somerville, Salem, Chelsea, Newton, Malden, Everett, Quincy, and Waltham, 42 percent;

Pittsburgh—the city, 66 percent, Allegheny (subsequently merged), 28 percent, and McKeesport, 6 percent;

New York—the city, 81 percent, and Newark, Jersey City, Paterson, Hoboken, Elizabeth, Passaic, Bayonne, Orange, West Hoboken, East Orange, and New Brunswick, 19 percent;

San Francisco—the city, 86 percent, and Oakland, 14 percent;

Philadelphia—the city, 91 percent, and Camden, Chester, and Norristown, 9 percent; and

Chicago—the city, 96 percent, and Joliet, Aurora, and Elgin, 4 percent.

Although exception may be taken to the inclusion of some specific outlying communities on grounds that their economic integration with the largest city was minimal, the combinations should lessen the impact of erratic annexation patterns. (An interesting sidelight on this point is the "disappearance" between 1850 and 1860 of the cities ranking, respectively, ninth, eleventh, twelfth, twentieth, and twenty-ninth with respect to population size at the Census of 1850; all were merged with Philadelphia, the fourth-ranking city of 1850.)

By comparison with the national industrial structure, the structures of the major centers are almost certain to reveal an overrepresentation of workers engaged in industries other than resource extraction. The traditional concentration of mercantile or distinctively metropolitan activities in large cities makes it equivocal, however, whether manufacturing workers will be overrepresented in the industrial structures of the major centers.

LOCALIZATION AND SPECIALIZATION

New York was the nation's foremost center of manufactures in 1900. Nearly 11 percent of the nation's manufacturing workers lived in New York, in contrast to the 5 percent living in Chicago, the second-ranking center of manufactures. Moreover, New York was preeminent in each of the six resource-market categories. The center's share of the national total was smallest with respect to the "second-stage, nonfinal" manufacturing industries, whose inputs were obtained from first-stage resource users and whose outputs were destined for other industrial establishments. Only 5 percent of the nation's workers employed in these nontraditional "manufacturers' manufactures" lived in New York. At the other extreme, nearly a fourth of the national work force employed in the "second-stage, final" manufacturing industries were New York residents.

From the perspective of localization, Chicago held a slight edge over Philadelphia as a center of manufactures; their respective shares of the national manufacturing work force were 5.0 and 4.7 percent. Localized in each center was at least 2.5 percent of the activity in each of the six resource-market categories. Chicago's share reached 9 percent among first-stage resource users producing for the final market; Philadelphia's share reached 8 percent among second-stage resource users producing for the final market. In an array of the major centers by their respective shares of the national total, the second rank was filled by Chicago or Philadelphia for all resource-market categories except the indirect-final; third rank was filled by Chicago or Philadelphia for all six market-resource categories.

No more than 2 percent of the national manufacturing work force were localized in any other center; seldom did the share of activity in any resource-market category reach 2.5 percent, the minimum observed in Philadelphia, Chicago, and New York. Boston ranked fourth among the major centers with respect to number of workers in all lines of manufacture, but Boston's share of the national total exceeded 2.5 percent only among manufacturers using resources indirectly, both those producing for the nonfinal market and those producing for the final market. Ranking lower with respect to number of workers in all lines of manufacture, but with noteworthy concentrations of activity in a specific resource-market category were: Baltimore, first-stage and second-stage resource users producing for the final market; and Cincinnati, manufacturers using resources indirectly and producing for the final market.

Together the sixteen largest population centers—ranging upward in size from New Orleans with 300,000 residents to New York with a population of four million—had localized within them a third of the nation's manufacturing workplaces. The large cities attracted relatively few of the manufacturers with a close dependence on resource extraction and a market orientation to other industrial establishments rather than households, however. Their aggregate share of the national total was just over a fifth among first-stage and second-stage resource users producing for the final market. (The just over a third among manufacturers using resources indirectly and producing for the nonfinal market. In contrast, their share was two-fifths among first-stage resource users producing for the final market, two-fifths among manufacturers using resources indirectly and producing for the final market, and nearly three-fifths among second-stage resource users producing for the final market. (The share of each center except Washington, in which manufacturing activity was negligible, is shown in Figure 5-1. The unshaded area within each diagram is proportional to the number of manufacturing workers living outside the major centers, the shaded area proportional to the number of manufacturing workers living within the centers.)

The historic concentration of households, the final market, toward the Atlantic seaboard and the westward diffusion of extractive activity are reflected in the shares of the Atlantic ports vis-à-vis the centers to their west. Boston, New York, Philadelphia, and Baltimore together had about as many manufacturing workers as did the twelve other major centers in four of the six resource-market categories, but the share of the Atlantic ports was more than twice the share of the remaining centers among the second-stage resource users producing for the final market and the manufacturers using resources indirectly and producing for the final market (Figure 5-1).

From the perspective of specialization, a rather different picture emerges. The preeminent manufacturing centers and the preeminent centers of manufacture were not in perfect correspondence.

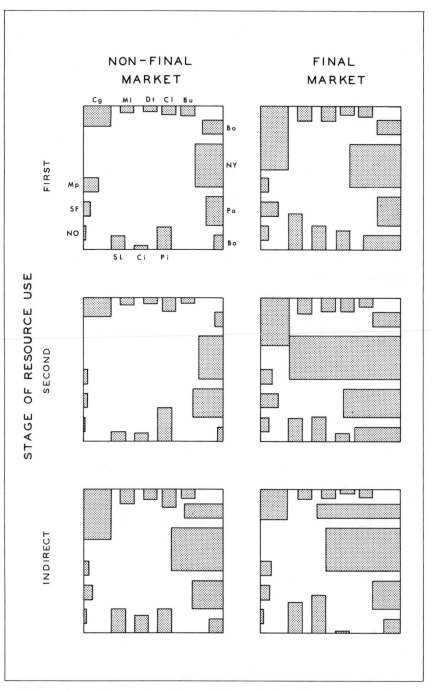

Figure 5-1. Percentage of national work force in major centers,
for six manufacturing industry categories: 1900.

It was in Cincinnati, Pittsburgh, Philadelphia, and Milwaukee that manufacturing workers were most heavily overrepresented in the local industrial structure although the largest numbers of manufacturing workers were found in New York, Chicago, and Philadelphia.

Urban manufactures had traditionally been in lines of industry producing for the final market of household consumers, such as processors of agricultural products or makers of apparel and footwear. Insofar as the manufactures had been an outgrowth of a city's position as metropolis for a resource region, the activity had a direct tie to resource extraction. The specialty common to the foremost manufacturing centers was, however, nontraditional manufactures —inputs heavily weighted with materials processed or fabricated by other manufacturers, outputs destined in large measure for other industrial establishments.

A classification of centers by the presence of each of three kinds of manufacture as a local specialty serves to highlight intercity differences in the resource-market orientation of the local industrial structure. Traditional metropolitan manufactures are taken to be represented by first-stage and second-stage resource users producing for the final market (1F, 2F). Traditional, but not reflecting a relation between metropolis and resource region are manufacturers producing for the final market and using resources indirectly (IF). Nontraditional are the second-stage users and manufacturers using resources indirectly who produce for the non-final market (2NF, INF). The center is "specialized" (S) in the given kind of manufacture if the local proportion in each component resource-market category, or average thereof, is at least as great as the corresponding proportion in the national nonextractive work force.

1F, 2F	IF	2NF, INF	
S	...	...	New Orleans, San Francisco
...	S	...	Boston
...	...	S	Pittsburgh, Cleveland
S	S	...	New York, Baltimore, St. Louis
S	...	S	Detroit
S	S	S	Philadelphia, Cincinnati, Chicago, Milwaukee

(The classification is developed from data presented in Table 5-1, in which the location quotient or ratio of the proportion of the

Table 5-1. Location Quotients in Six Manufacturing Industry Categories Defined by Stage of Resource Use and Type of Market, for Major Centers, 1900

Center	Symbol	Manufacturing, All	FIRST STAGE Nonfinal	FIRST STAGE Final	SECOND STAGE Nonfinal	SECOND STAGE Final	INDIRECT Nonfinal	INDIRECT Final
Boston	Bo	1.27	1.01	1.15	0.41	1.36	1.90	4.05
New York	NY	1.74	0.95	1.79	0.85	3.90	1.81	2.63
Philadelphia	Pa	2.21	1.18	1.63	2.10	3.80	2.06	1.93
Baltimore	Ba	1.75	0.80	3.61	0.60	4.44	1.43	1.74
Cincinnati	Ci	2.45	0.79	3.40	1.24	3.75	2.64	5.74
St. Louis	SL	1.69	1.13	2.94	0.87	2.11	2.09	2.62
Chicago	Cg	1.99	1.13	3.54	1.06	2.67	2.78	1.70
Pittsburgh	Pi	2.39	2.33	1.89	3.48	0.83	2.50	0.29
Buffalo	Bu	1.31	1.64	1.72	0.96	1.65	1.43	1.23
Cleveland	Cl	1.93	1.22	1.19	2.03	2.08	2.62	0.74
Detroit	Dt	1.86	1.09	2.86	1.36	2.41	2.21	1.65
Milwaukee	Ml	2.20	1.55	3.10	1.56	2.93	2.64	2.20
New Orleans	NO	0.80	0.48	1.68	0.42	1.78	0.52	0.77
San Francisco	SF	1.13	0.98	2.06	0.61	1.96	1.08	1.18
Washington	Wa	0.43	0.80	0.48	0.22	0.52	0.64	0.32
Minneapolis	Mp	1.12	2.21	1.11	0.57	1.65	1.23	1.57
Kansas City*	KC	1.18	0.98	5.48	0.45	1.16	0.69	0.65
Los Angeles*	LA	0.77	1.27	1.24	0.58	0.85	0.68	0.76

* Displaced New Orleans, smallest of the major centers, in rank with respect to population size between 1900 and 1920.

local work force engaged in manufactures to the proportion of the national work force engaged in manufactures is shown for each center and resource-market category. The local proportion is as great as the proportion in the national work force exclusive of the extractive sector when the location quotient takes on a value of 1.7 or more).

It was in New Orleans and San Francisco, isolated from the main line of settlement and the manufacturing belt, proximate to resource regions recently opened to extractive activity, that the sole manufacturing specialty was a direct outgrowth of the center's role as a regional metropolis. This traditional-metropolitan manufacturing specialty appeared in a number of other major population centers, but always in conjunction with another kind of manufacturing specialty. Moreover, in very few major centers was the proportion of the local work force engaged in all branches of manufacture as low as in New Orleans and San Francisco (Table 5-1).

If the proportion of the local work force engaged in manufactures was no greater than the proportion in New Orleans and San Francisco, a major population center probably should be excluded from the roster of manufacturing centers on grounds that manufacturing is an ancillary rather than primary local function. The exclusions would number two, Washington and Minneapolis. In the nation's capital, all kinds of manufacture were underrepresented not only by comparison with the national nonextractive work force, but by comparison with the total national work force as well. The sole specialty in Minneapolis, the new rail gateway to the northwest, was in lines of manufacture classified as first-stage resource users producing for the nonfinal market. Inasmuch as this atypical large-city specialty reflected a concentration of grain milling in Minneapolis, it can be regarded as an extension of the center's role as metropolis for the mid-continent wheat belt. (The same kind of specialization in Pittsburgh is less clearly traditional and metropolitan, for it reflects a local concentration of marble and stone work.)

The roster of manufacturing centers included seven cities in the northeast which had attained national prominence before the industrialization of manufactures—Boston, New York, Philadelphia, Baltimore, Cincinnati, St. Louis, Chicago—and five new centers encircled by these established metropolises—Pittsburgh, Buffalo, Cleveland, Detroit, and Milwaukee. The traditional-metropolitan specialty was present in all the older centers in which manufacturing had become a primary function except Boston; it was absent in all the new centers except Detroit and Milwaukee. Traditional, but not distinctively metropolitan manufactures were a specialty of each older center, no new center save Milwaukee. Only three older centers, Philadelphia, Cincinnati, and Chicago, had developed nontraditional manufactures as a local specialty; but the nontraditional specialty appeared in all new centers save Buffalo.

The regional metropolises had not lost the kinds of manufacture that a metropolis traditionally performed for resource extractors and household consumers in its tributary territory although relocation of manufacturing enterprise continued to occur as the resources of some regions were exhausted, the resources of others opened to exploitation. It was not an accelerated redistribution of traditional-metropolitan manufactures, however, that heightened differentiation among major population centers with respect to the amount of manufacturing activity in the local economy shortly before 1900. Rather, industrial manufactures, especially nontraditional branches such as iron and steel and their products, had become the national growth industry. Some metropolises were more successful than others in attracting a share of the new manufactures which was commensurate with the scale of their commercial base or the strength of their financial function. Some lesser centers captured substantially more than their "fair" shares of the growth industry and were, thereby, transformed into major population centers, if not metropolises.

THE INDUSTRIAL PROFILES

The fortunes of the manufacturing centers were not tied to the performance of establishments in a single branch of manufacture. Philadelphia and Cincinnati, for example, not only had a substantial overrepresentation of workers in each of the three kinds of manufacture distinguished on the basis of resource-market orientation; in their local industrial structures were some twenty-five branches of manufacture which would be identified as specialties on the rather stringent criterion that the proportion of local workers engaged in the industry was at least double the corresponding proportion in the national work force. Boston and Pittsburgh, the manufacturing centers specialized in a single kind of manufacture, were least diversified in the manufacturing sector; but their industrial profiles included fifteen and thirteen detailed manufacturing specialties, respectively. Only in the large cities excluded from the roster of manufacturing centers did local industrial specialties in the manufacturing sector number fewer than a dozen.

Although all the manufacturing centers had a diversity of detailed manufacturing specialties, their local manufactures tended to have a distinctive character, which becomes popularized in such images as Chicago the hog butcher or Pittsburgh the steel maker. There were in each center a relatively few industries which were both local specialties and major local employers. The meat-products industry employed 4 percent of Chicago's workers, fourteen times the national proportion of workers in the industry. Blast furnaces, steel works, rolling and finishing mills providing employment for 18 percent of the Pittsburgh work force, twenty times the national proportion. These are, to be sure, distinguishing features of the respective local economies.

Identification of centers in terms of such distinguishing features easily conveys the impression of a one-industry center which is the site of the nation's largest concentration of activity in that industry. The impression is likely to be false, however. With this caution, detailed industrial profiles in the manufacturing sector are presented for each major center; and certain distinguishing features are singled out for discussion. In the tabular presentation, the detailed industrial specialties in the manufacturing sector are grouped by resource-market orientation; and those industries which are both local specialties (a location quotient of 2.0) and major local employers (one percent of the local work force) are denoted by their underscored cell entries.

Relatively few first-stage resource users were prominent in the industrial profiles of the seven northeastern centers which had established themselves as regional metropolises before the industrialization of manufactures (Table 5-2). That such specialties which depend on an access advantage to sites of natural resource extraction are especially vulnerable to change or transitory in nature is clear. Chicago, in 1900, was a slaughtering and meat-packing center. The eclipse of Cincinnati by Chicago, as described by the Census statisticians of 1900, only points up the instability in Chicago's position, however.

Slaughtering and meat packing, as the industry is now understood, had its beginning at Cincinnati, Ohio, about 1818. Since that time

Table 5-2. Industrial Profiles in the Manufacturing Sector for Boston, New York, Philadelphia, Baltimore, Cincinnati, St. Louis, and Chicago, 1900

Stage of Resource-Use, Type of Market, and Detailed Industry	LOCATION QUOTIENT FOR						
	Bo	NY	Pa	Ba	Ci	SL	Cg
First Stage, Nonfinal							
Misc. nonmetallic mineral and stone products	3.1	...ᵃ	...	2.2	...	...	...
Petroleum refining	...	3.1	3.2	...	...	...	2.4
First Stage, Final							
Confectionary and related products	4.2	2.8	3.6	3.7	3.1	3.8	3.1
Tobacco manufactures	...	2.7ᵇ	2.1	3.7	5.5	3.9	...
Beverage industries	...	2.4	...	...	5.4	5.6	2.2
Misc. food preparations and kindred products	...	...	2.0	...	...	2.5	2.3
Canning and preserving fruits, vegetables, and seafoods	...	...	...	12.4	...	...	...
Meat products	...	...	...	...	2.5	...	14.3
Second Stage, Nonfinal							
Paints, varnishes, and related products	...	4.7	5.4	...	6.0	5.0	4.0
Misc. textile mill products	...	...	4.8	2.8	...	...	...
Yarn, thread, and fabric mills	...	...	3.2	...	...	...	...
Primary iron and steel industries, except blast furnaces, etc.*	...	...	2.9	...	3.9	...	2.6
Cement and concrete, gypsum, and plaster products	...	...	2.8	...	...	3.2	...
Misc. petroleum and coal products	...	...	...	...	2.3	2.8	...
Second Stage, Final							
Bakery products	2.4	3.2	2.9	3.2	3.7	2.9	2.9
Misc. fabricated textiles	...	7.5	2.2	2.7	5.3	...	2.9
Apparel and accessories	...	5.5	2.9	6.9	4.2	2.4	3.1
Floor coverings, except hard surface	...	...	18.9	...	...	...	...
Knitting mills	...	...	7.1	...	...	...	...
Furniture and fixtures	...	...	...	2.3	5.0	2.9	3.4
Indirect, Nonfinal							
Misc. machinery	4.9	...	9.4	...	...	5.5	4.2
Printing, publishing, and allied industries, except newspapers	4.3	4.2	3.7	2.2	5.5	3.6	4.9
Rubber products	3.8	...	...	...	...	...	...

Table 5-2. (continued)

| Stage of Resource-Use, Type | | | LOCATION | QUOTIENT | FOR | | |
of Market, and Detailed Industry	Bo	NY	Pa	Ba	Ci	SL	Cg
Indirect, Nonfinal—cont.							
Electrical machinery, equipment, and supplies	*3.7*	2.2	...	...	...	...	3.9
Drugs and medicines	2.6	2.3	4.1	6.1	3.9	5.3	2.6
Paperboard containers and boxes	2.5	3.7	4.5	...	4.0	2.4	3.4
Newspaper publishing and printing	*2.4*	*2.1*	...	...	2.3	...	...
Professional equipment and supplies	2.3	2.2	7.9	...	2.6	...	2.5
Leather: tanned, curried, and finished	2.2	...	*5.4*	...	3.6	...	...
Misc. fabricated metal products	2.1	2.4	2.0	*5.6*	*4.2*	...	...
Misc. paper and pulp products	...	4.8	2.6	...	2.9	...	...
Fabricated structural metal products	...	3.7	2.0	...	3.0	4.2	4.7
Ship and boat building and repairing	...	...	...	4.0	...	...	...
Cutlery, hand tools, and other hardware	...	...	2.3	...	2.1	...	...
Railroad and misc. transportation equipment	...	...	...	...	*4.3*	*4.2*	3.8
Primary nonferrous industries	...	...	...	...	2.0	...	...
Pottery and related products	...	...	...	...	...	3.4	...
Farm machinery	...	...	...	...	...	...	*9.2*
Indirect, Final							
Footwear, except rubber	*7.5*	...	...	...	*9.9*	*4.5*	...
Misc. manufacturing industries	*2.0*	*4.2*	*2.4*	*2.2*	*2.9*	...	*2.3*
Watches, clocks, and clockwork-operated devices	...	2.6	...	...	...	...	...
Leather products, except footwear	...	2.3	...	...	2.9	2.3	2.1

a. .... indicates location quotient less than 2.0, or less than 0.015 percent of national work force in industry.

b. Italic figures indicate at least 1.0 percent of local work force in industry.

* Includes "Iron and steel and their products: Foundry and machine shop products."

the center of the industry has moved gradually westward, following the development of new cattle and swine producing sections. This tendency has been intensified by the perfection of artificial refrigeration and refrigerator cars, which has made the difference between the cost of transporting live stock, and meat as a finished product, sufficient to induce packers to establish plants near the stock-raising or stock-fattening sections. These sections, in turn, are determined by the production of grain, principally corn and hay, so that the localization of the packing industry is influenced in a large degree by the production of these agricultural staples. [Twelfth Census, Vol. 7, Pt. 1: ccviii]

Not only is the depletion of resources at the center's existing supply sites a risk. The discovery of new and richer sites, the introduction of new methods of extraction or processing, innovations in transporting raw material or product can equally well wipe out a center's competitive advantage.

The most common feature of manufactures in these older centers which reflects unambiguously a center-region tie was the tobacco specialty. The south Atlantic district was an important domestic producer of cigarette tobacco, Pennsylvania of cigar filler; and employment in New York, Philadelphia, and Baltimore was more heavily concentrated in the cigars-and-cigarette branch of the industry than in the manufacture of chewing-and-smoking tobacco and snuff. Tennessee, Kentucky, and West Virginia were producers of the Burley tobacco used by manufacturers of chewing-and-pipe tobacco; and in Cincinnati and St. Louis, the cigars-and-cigarettes branch was less important than in New York, Philadelphia, and Baltimore. Neither Boston nor Chicago, more remote from the tobacco-growing areas, had developed the specialty.

Brewing was a first-stage specialty of Cincinnati and St. Louis, centers relatively near cereal-producing areas and including among their residents large numbers of Germans accustomed to quality beer and skilled in the art of brewing. Baltimore alone specialized in the canning industry, which in Maryland had developed around the processing of tomatoes and corn.

Some first-stage specialties found in the newer manufacturing centers were almost certain to wane. Buffalo and Milwaukee, for example, had as a specialty the products of planing mills; but lumbering in the surrounding territory had been decreasing as the forest resources of New York State and Wisconsin neared depletion. Less clearly vulnerable were the marble and stone work and the preparation of pickles, preserves, and sauces in Pittsburgh, the manufacture of cigars and cigarettes in both Pittsburgh and Detroit, and the brewing of malt liquors in Milwaukee (Table 5-3).

Table 5-3. Industrial Profiles in the Manufacturing Sector for Pittsburgh, Buffalo, Cleveland, Detroit, and Milwaukee, 1900

Stage of Resource-Use, Type of Market, and Detailed Industry	LOCATION QUOTIENT FOR				
	Pi	Bu	Cl	Dt	Ml
First Stage, Nonfinal					
Misc. nonmetallic mineral and stone products	8.2[b]	...[a]	...	...	...
Petroleum refining	...	...	2.8	...	...
Sawmills, planing mills, and millwork	...	3.8	...	2.9	3.6
First Stage, Final					
Confectionary and related products	2.4	3.3	2.4	3.0	3.4
Tobacco manufactures	2.1	...	...	6.1	...
Beverage industries	...	3.4	...	2.4	12.0
Canning and preserving fruits, vegetables, and seafoods	4.7	...	...	...	...
Meat products	...	3.2	...	...	2.8
Second Stage, Nonfinal					
Paints, varnishes, and related products	3.6	2.8	9.5	13.3	...
Misc. textile mill products	...	...	2.4	...	...
Primary iron and steel industries, except blast furnaces, etc.*	3.7	2.2	5.0	4.2	5.4
Cement and concrete, gypsum, and plaster products	...	2.6	2.4	2.1	...
Misc. petroleum and coal products	2.4	2.4	...	2.2	...
Blast furnaces, steel works, and rolling and finishing mills	19.9	...	5.2	...	...
Misc. chemicals and allied products	...	2.6	...	...	...
Second Stage, Final					
Bakery products	3.0	3.3	...	2.6	3.3
Misc. fabricated textiles	...	...	...	...	2.6
Apparel and accessories	...	...	2.7	2.5	2.6
Knitting mills	...	...	...	...	4.4
Furniture and fixtures	...	2.5	...	3.7	3.3
Indirect, Nonfinal					
Printing, publishing, and allied industries, except newspapers	...	3.4	2.5	3.0	3.7
Rubber products	...	...	3.4	...	...
Electrical machinery, equipment, and supplies	11.5	...	10.2	...	...
Drugs and medicines	...	4.7	...	24.1	...
Paperboard containers and boxes	...	2.9	2.1	3.4	...

Table 5-3. (continued)

Stage of Resource-Use, Type of Market, and Detailed Industry	LOCATION QUOTIENT FOR				
	Pi	Bu	Cl	Dt	Ml
Indirect, Nonfinal—cont.					
Newspaper publishing and printing	...	...	...	...	2.2
Professional equipment and supplies	...	...	...	3.7	...
Leather: tanned, curried, and finished	...	2.2	...	...	14.1
Misc. fabricated metal products	...	...	7.2	7.4	6.0
Fabricated structural metal products	14.4	...	4.2	5.2	4.9
Cutlery, hand tools, and other hardware	...	...	7.6	...	...
Railroad and misc. transportation equipment	...	...	...	...	2.1
Primary nonferrous industries	...	...	...	4.5	2.5
Pottery and related products	6.6	...	...	...	...
Farm machinery	...	...	...	...	4.9
Glass and glass products	5.5	...	...	...	...
Indirect, Final					
Footwear, except rubber	...	...	...	...	2.8
Misc. manufacturing industries	...	...	...	2.3	...
Leather products, except footwear	...	2.4	...	...	4.5

a. ... indicates location quotient less than 2.0, or less than 0.015 percent of national work force in industry.
b. Italic figures indicate at least 1.0 percent of local work force in industry.
* Includes "Iron and steel and their products: Foundry and machine shop products."

The regional metropolises of New Orleans, San Francisco, and Minneapolis each had a first-stage specialty which was closely related to agricultural activity in a nearby resource region (Table 5-4). The New Orleans specialty in the food-preparations industry reflected a local concentration of sugar and molasses refining which drew upon both imports of raw sugar and supplies from the lower Mississippi Valley where the domestic production of sugar cane was concentrated. San Francisco specialized in canning and preserving, and an important component of the industry in California was dried-fruit products, particularly prunes and raisins. Prominent in the Minneapolis profile was the flouring and grist-mill industry, in which the leading product was white wheat flour.

The other manufacturing specialties found in the major centers are, of course, those in which the relation between manufacturer and resource extractor or center and resource region is less direct.

Table 5-4. Industrial Profiles in the Manufacturing Sector for New Orleans, San Francisco, Washington, and Minneapolis, 1900

Stage of Resource-Use, Type of Market, and Detailed Industry	LOCATION QUOTIENT FOR			
	NO	SF	Wa	Mp
First Stage, Nonfinal				
Sawmills, planing mills, and millwork	...[a]	...	...	2.6
Grain-mill products	...	...	...	5.1[b]
First Stage, Final				
Confectionary and related products	...	2.6	...	2.3
Beverage industries	2.1	...	...	...
Misc. food preparations and kindred products	4.8	3.2	...	...
Canning and preserving fruits, vegetables, and seafoods	...	4.5	...	...
Second Stage, Nonfinal				
Cement and concrete, gypsum, and plaster products	...	3.5	4.4	...
Misc. petroleum and coal products	2.7	...	...	...
Second Stage, Final				
Bakery products	3.4	2.2	2.4	2.2
Misc. fabricated textiles	3.1	2.4	...	...
Apparel and accessories	2.1	2.7	...	...
Indirect, Nonfinal				
Misc. machinery	...	2.0	...	...
Printing, publishing, and allied industries, except newspapers	...	2.6	...	2.4
Drugs and medicines	2.1	...	...	2.0
Newspaper publishing and printing	...	2.1	...	3.3
Fabricated structural metal products	...	...	...	3.9
Indirect, Final				
Footwear, except rubber	...	...	...	2.3
Leather products, except footwear	...	...	...	2.7

a. ... indicates location quotient less than 2.0, or less than 0.015 percent of national work force in industry.

b. Italic figures indicate at least 1.0 percent of local work force in industry.

It is these specialties which are potentially persistent and with respect to which an association between "age" of center and "age" of specialty should be discernible if differentiation among centers resulted from the distribution of new lines of manufacture rather than the redistribution of old lines.

The twelve centers with a well-developed industrial profile in the manufacturing sector can be ordered, at least approximately, by age. Boston, New York, Philadelphia, and Baltimore are unquestionably older than Cincinnati, St. Louis, and Chicago which, in turn, are older than Pittsburgh, Buffalo, Cleveland, Detroit, and Milwaukee. When the specialties are identified by name and ordered by the increasing age of the center(s) in which the specialty appears, the specialties listed first "sound" newer than those which follow (Table 5-5). New and old lines of manufacture cannot be distinguished systematically, for time series on the number of wage earners in each industry which appears as a specialty are not available.

Table 5-5. Centers in Which a Given Industry Was Both a Local Specialty and a Major Local Employer in 1900

Resource-Use and Market-Type Category and Title of Component Industry Employing Largest Number of Workers	Center
Second Stage or Indirect, Nonfinal	
Bicycles and tricycles	— — — — — — — — — — — Ml
Steam fittings and heating apparatus	— — — — — — — — — — Dt —
Druggists' preparations	— — — — — — — — — — Dt —
Bolts, nuts, washers, and rivets	— — — — — — — — — Cl — —
Hardware	— — — — — — — — — Cl — —
Iron and steel	— — — — — — — Pi — Cl — —
Soap and candles	— — — — — — — — Bu — — —
Architectural and ornamental iron work	— — — — — — — Pi — — — —
Pottery, terra cotta, and fire clay products	— — — — — — — Pi — — — —
Glass	— — — — — — — Pi — — — —
Agricultural implements	— — — — — — Cg — — — — Ml
Steam-railroad cars	— — — — — — Cg — — — — —

Table 5-5. (continued)

Resource-Use and Market-Type Category and Title of Component Industry Employing Largest Number of Workers	Center											
Carriages and wagons	—	—	—	—	Ci	SL	—	—	—	—	—	—
Safes and vaults	—	—	—	—	Ci	—	—	—	—	—	—	—
Foundry and machine shop products	—	—	—	Pa	Ci	—	Cg	Pi	Bu	Cl	Dt	Ml
Electrical apparatus and supplies	Bo	—	—	—	—	—	Cg	Pi	—	Cl	—	—
Printing and publishing, book and job	Bo	NY	Ba	Pa	Ci	SL	Cg	—	Bu	Cl	Dt	Ml
Newspaper printing and publishing	Bo	NY	—	—	Ci	—	—	—	—	—	—	Ml
Enameling and enameled goods	—	—	Ba	—	—	—	—	—	—	—	—	Ml
Leather: tanned, curried, and finished	—	—	—	Pa	—	—	—	—	—	—	—	Ml
Cotton goods	—	—	—	Pa	—	—	—	—	—	—	—	—
Second Stage or Indirect, Final												
Matches	—	—	—	—	—	—	—	—	—	—	Dt	—
Furniture	—	—	—	—	Ci	SL	Cg	—	Bu	—	Dt	Ml
Coffins, burial cases, etc.	—	—	—	—	Ci	—	—	—	—	—	—	—
Bread and other bakery products	—	—	—	—	Ci	—	—	—	—	—	—	—
Men's clothing	—	—	Ba	Pa	Ci	SL	Cg	—	—	—	Dt	Ml
Boots and shoes	Bo	—	—	—	Ci	SL	—	—	—	—	—	Ml
Hosiery and knit goods	—	—	—	Pa	—	—	—	—	—	—	—	Ml
Women's clothing	—	NY	—	—	—	—	—	—	—	Cl	—	—
Musical instruments, pianos	Bo	NY	—	—	—	Cg	—	—	—	—	—	—
Umbrellas and canes	—	—	Ba	Pa	—	—	—	—	—	—	—	—
Carpets and rugs, other than rag	—	—	—	Pa	—	—	—	—	—	—	—	—

Note: Industries appear with underscored cell entries in Tables 5-2 and 5-3. Identifying component may differ among centers, e.g., bicycles and tricycles, steam-railroad cars, and carriages and wagons for railroad and miscellaneous transportation equipment. Full identification of cell symbols appears in Table 5-1.

It is possible to contrast with respect to age the specialties which appear solely in Milwaukee and Detroit, the youngest cities, and the specialties found only in one of the old Atlantic ports. The specialties of the youngest cities were bicycles and tricycles, steam fittings and heating apparatus, and druggists' preparations. Wage earners in these industries in the nation as a whole have been reported as: bicycles and tricycles, 18,000 workers in 1900, 2,000 in

1890; steam fittings and heating apparatus, 9,000 in 1900, 2,000 in 1880; and druggists' preparations, 6,000 workers in 1900, 2,000 in 1890. Cotton goods, umbrellas and canes, and carpets and rugs distinguished the manufacturing sector in the old seaports. The national work force in these industries had numbered: cotton goods, 303,000 in 1900, 122,000 in 1860; umbrellas and canes, 6,000 in 1900, 2,000 in 1860; and carpets and rugs, 28,000 in 1900, 7,000 in 1860. The association between the "ages" of center and specialty is by no means as clear-cut as this contrast would suggest, for in a number of instances the same specialty appears in both young and old centers.

In at least one instance, a specialty as pursued in a young center differed significantly from what was ostensibly the same specialty in an old center. The industry, leather: tanned, curried, and finished, was unambiguously old; yet it appeared as a specialty in the profile of Milwaukee as well as in Philadelphia. Details of this specialization can be documented quite fully, because the industry was a subject covered by a special report for the states of Wisconsin and Pennsylvania in the Thirteenth Census (Vol. 9). Shown below are the materials used by the industry in the respective States in the year 1899:

Millions of—	*Wisconsin*	*Pennsylvania*
Hides	2.1	4.8
Skins:		
Calf and kip	2.8	0.4
Goat	0.2	21.9
Other		2.0

Demand for calf and kip skins presumably was met from domestic sources, for in 1909 some 5.1 million skins were reported taken from calves in the United States. Imports were the major source of goatskins, however; for only 0.3 million skins were reported taken from goats in the United States. (Statistics on number of hides and skins taken off animals were reported in the Thirteenth Census, Vol. 8, Manufactures: 429.) Tanbark was then abundant as a by-product of the lumbering industry in Wisconsin. Although hemlock still was present in the forests of northern and central Pennsylvania, as early as 1860 tanning by the use of chemicals had been perfected

in Philadelphia (Twelfth Census, Vol. 8, Manufactures: 751).
Whether this is an isolated instance remains moot.

Some specialties similar in all significant aspects will appear
in both a young center and an old center if differentiation among
centers arises in minor part from a redistribution of traditional
manufactures and in major part from a distribution of new lines of
manufacture such that a center's share of the growth industry is
not simply proportional to the size of its existing manufacturing
base. In fact, had an old center failed to develop any new specialties, it could not long have maintained its position among the ranks
of the major centers. Nonetheless, there is ample evidence of a relation between the recency of growth in a center and the relative
importance of new lines of activity in the local industrial profile.

Chapter **6**

THE METROPOLITAN FINANCIAL FUNCTION IN 1900

Reserve-city status in the national banking system serves to identify places whose banks play a strategic role in the organization of the system. Shifts in status can be traced forward from 1864, so that the first measurement antedates the peopling of the trans-Mississippi region as well as the industrialization of manufactures.

Through one means or another, the monetary systems of most nations restrict the proportion of deposits which a bank may loan or otherwise invest. Banks are required by law or prudence to keep a portion of their funds on reserve. In the United States, the Federal Reserve Board currently determines bank reserve requirements and may vary the proportions in accordance with the state of the economy and the Board's desire to retard or expand the availability of credit (Board of Governors of the Federal Reserve System, 1938: 953); but laws requiring certain fixed reserves existed long before the Federal Reserve Act of 1913.

The National Bank Act of 1863 required not only that national banks maintain a reserve against their notes and deposits, but named nine cities as "redemption centers." When the law was revised in the following year, seventeen cities were given this special classification. In effect, a banking hierarchy with three distinct layers was created. Occupying the topmost position was New York, the only central-reserve city. The next layer included the sixteen remaining redemption centers or, as they were later called, "reserve cities." Banks located in all other places were at the bottom of the pyramid. These became known as "country banks," a term which applies to all banks, whether in big cities or the smallest hamlet, that are located outside a reserve or central-reserve city.

There were unique advantages as well as risks inherent in each of the three levels. Country banks were required to maintain a reserve of 15 percent against their notes and deposits; three-fifths of this reserve could be kept in interest-bearing deposits with the national banks of any of the seventeen redemption cities. Banks in the reserve cities were required to maintain a reserve of 25 percent, but these cities could more than compensate for the higher reserve requirements by attracting sufficient deposits from the country banks. Moreover, the sixteen ordinary reserve cities were allowed to carry half their required reserves in approved banks in New York City and gain interest on the money. Reserve status could be quite profitable despite the higher reserve requirements if the reserve-city banks could capture the reserve deposits of a number of country banks. New York's banks faced tighter restrictions since they had to keep a full reserve of 25 percent in their own vaults. The loss of interest could be compensated for, however, by the fact that only the New York banks were eligible to hold the reserve deposits of the other redemption centers.

Rodkey (1934: 384) has described the original rationale for the naming of reserve or redemption cities:

> It seemed essential, in order to avoid a discounted currency, that banks be required to redeem their notes at other places than over their own counters. For this purpose redemption cities were named and banks were required to provide for the redemption of their notes in one or more of these cities. The required reserve was considered only secondarily as a safety fund for depositors; primarily it was intended to insure the convertibility of notes into specie. Since it was necessary for banks to maintain specie with their redemption agents in order that their notes might be redeemed, and since this was thought of as the most important purpose back of the fixed specie reserve requirement, it seemed only reasonable to allow these banks to count their balances with redemption agents as part of their legal reserve.

The banking laws were amended in 1887 to allow more places to become reserve cities and to permit additional central-reserve cities. Upon the request of three-fourths of the national banks in any city with at least 200,000 residents, central-reserve status could be obtained. A request by three-fourths of the national banks in any city with at least 50,000 residents could provide for reserve-city status. The minimum population size for reserve cities was reduced to 25,000 in 1903 (Board of Governors of the Federal Reserve System, 1938: 961); and by 1910 reserve cities in the national banking system numbered forty-seven.

In summary, the classification of cities into central-reserve, reserve, and country-bank statuses provides a set of discrete categories which may be used to trace out, on a crude basis, the banking significance of various cities over time. The added restrictions as well as the potential gains of reserve cities made attainment and maintenance of this status a reflection of the city's ability to tap a financial hinterland. Movement between these discrete categories provides, then, an index for tracing competition between aspiring cities in newer regions as well as the rise and fall of banking centers in older parts of the nation.

RESERVE CITIES

Both population size and distance from competing centers influenced whether a city attained reserve status as well as whether it maintained its special position. In 1860, the nation's largest city had been New York, which was designated the central-reserve city. Next in rank had been Philadelphia, Baltimore, Boston, and New Orleans. These cities as well as St. Louis, Chicago, and Cincinnati, which were to displace New Orleans in rank with respect to population size during the 1860s, were designated redemption centers. In fact, of the twenty-one largest cities in 1860, all but five were named redemption centers under the National Bank Act of 1864. The five excluded cities probably were overshadowed by New York City. They were: Brooklyn (to be consolidated with New York in 1898, but then the third-largest city in the nation) and Newark, both in the immediate environs of the City; Buffalo and Rochester in upstate New York, not far from the City, even closer to redemption centers such as Albany and Cleveland; and Providence, Rhode Island which lay between New York and Boston, a reserve city.

Leavenworth, Kansas was the only redemption center established in 1864 which did not rank among the top twenty-one cities. Its population in 1860 was only 7,000 and, although the population was to more than double by 1870, the city was far too small to be nationally prominent, ranking about one hundredth among the nation's cities. Nevertheless, Leavenworth in eastern Kansas was by far the largest city in the state and also was bigger in 1860 than neighboring Kansas City, Missouri. Leavenworth faced no real competition for a long distance to its west, being larger in 1860 than Denver, Colorado and nearly as big as Salt Lake City, Utah. Likewise, it was prominent relative to places to its north; Omaha, Nebraska in 1860 had fewer than 2,000 residents, and neither Sioux City, Iowa nor Sioux Falls, South Dakota were then competitors. One had to journey to New Orleans, or to Galveston or San Antonio in Texas to reach a city larger than Leavenworth to its south. The only nearby competitor was St. Joseph, Missouri, a town of 9,000. Thus in 1860, among cities west of St. Louis, Leavenworth enjoyed a good position to carve out a banking hinterland in a vast and relatively undeveloped area. With the exception of St. Joseph, an urban vacuum existed in the sparsely populated areas to Leavenworth's north, south, and west. While one could hardly claim a set of ecological principles that would make selection of Leavenworth under the National Banking Act inevitable, the city's inclusion can be understood in terms of the proposition that holding city size constant, distance from larger centers will facilitate the development of metropolitan functions. (On this point see, for example, Duncan et al., 1960: 127–128).

Leavenworth's reserve-city status was terminated in 1872. The demise of Leavenworth can be understood in the same terms as its inclusion earlier. In the ensuing decades Leavenworth failed to grow as rapidly as some of the neighboring cities which soon surpassed it by enormous margins. Since Leavenworth fell behind in rate of growth at a point when its own population was hardly sufficient for metropolitan status, it was extremely vulnerable. Between 1860 and 1870, Leavenworth more than doubled in population and had nearly 18,000 residents by the end of the decade;

no further gain was recorded in the 1870s. During the same period, nearby Kansas City leaped from 4,000 to 32,000 to 56,000; Omaha expanded from 2,000 to 16,000 to 31,000; and St. Joseph increased from 9,000 to 20,000 to 32,000.

Not until 1887 were any new reserve cities created. The first three additions were located on the mid-continent plain: namely, Kansas City; Omaha (Nebraska); and St. Joseph (Missouri). By 1900 an additional twelve cities had been added to the reserve-status category by the Comptroller, the vast majority located west of the Mississippi. Only Brooklyn, Savannah (Georgia), Columbus (Ohio), and Indianapolis (Indiana) were added from the territory east of the Mississippi. To the west, the new reserve cities included: Minneapolis and St. Paul, Des Moines (Iowa), Lincoln (Nebraska), and Houston (Texas) in mid-continent; Denver (Colorado) in the Mountain States; and Los Angeles and Portland (Oregon) on the Pacific seaboard (see Table 6-1).

Holding reserve status in 1900 were fifteen cities which had been designated redemption centers in 1864 and remained among the nation's twenty-five largest cities at the turn of the century. Still without reserve status in 1900 were four large cities which had not been designated redemption centers in 1864—Newark, Buffalo, Rochester, and Providence. Among the cities that had shifted upward in rank with respect to population size into the "top twenty-five," Indianapolis, Minneapolis, St. Paul, Kansas City, and Denver had become reserve cities; Jersey City, a near neighbor of New York, had not. The same forces which had operated initially to block the development of financial importance among some of the larger eastern cities seemed to continue in the late 1800s.

The influences of city size and proximity to competing centers also can be observed in the creation of central-reserve cities. Chicago, the nation's third-largest population center in 1880, had fought for the 1887 amendment permitting the creation of additional central-reserve cities on the grounds that its banks could compete successfully with New York banks for the deposits of reserve-city banks. This inflow would more than compensate for the fact that half of Chicago's own reserve requirements no longer could earn interest by being deposited in New York banks, but must, instead, be stored in the vaults of the Chicago banks. Neither Philadelphia, the nation's second-largest center, nor Boston, the nation's fourth-

Table 6-1. Reserve Cities in the National Banking System from 1864 to 1910 (x indica reserve-city status)

Region and Center	1864	1887	1900	1910
Atlantic Seaboard				
Boston	x	x	x	x
New York City[a]	x	x	x	x
Philadelphia	x	x	x	x
Baltimore	x	x	x	x
Washington	x	x	x	x
Albany, N.Y.	x	x	x	x
Savannah, Ga.	...	...	x	x
Transmontane East				
New Orleans	x	x	x	x
Cincinnati	x	x	x	x
Chicago	x	x	x	x
Pittsburgh	x	x	x	x
Cleveland	x	x	x	x
Detroit	x	x	x	x
Milwaukee	x	x	x	x
Louisville, Ky.	x	x	x	x
Columbus, Ohio	...	...	x	x
Indianapolis, Ind.	...	...	x	x
Mid-Continent				
St. Louis[b]	x	x	x	x
Minneapolis[c]	...	...	x	x
Kansas City[d]	...	x	x	x
Leavenworth, Kans.	x	...	...	...
Omaha, Neb.	...	x	x	x
St. Joseph, Mo.	...	x	x	x

Region and Center	1864	1887	1900
Mid-Continent—cont.			
Des Moines, Iowa	...	...	x
Lincoln, Neb.	...	...	x
Houston, Texas	...	...	x
Denver, Colo.	...	...	x
Wichita, Kans.	...	...	...
Topeka, Kans.	...	...	...
Cedar Rapids, Iowa	...	...	...
Dubuque, Iowa	...	...	...
Sioux City, Iowa	...	...	...
So. Omaha, Neb.	...	...	...
Muskogee, Okla.	...	...	...
Oklahoma City	...	...	...
Dallas, Texas	...	...	...
Fort Worth, Texas	...	...	...
Galveston, Texas	...	...	...
San Antonio, Texas	...	...	...
Waco, Texas	...	...	...
Pueblo, Colo.	...	...	...
Far West			
San Francisco	x	x	x
Los Angeles	...	...	x
Portland, Ore.	...	...	x
Seattle, Wash.	...	...	...
Spokane, Wash.	...	...	...
Tacoma, Wash.	...	...	...
Salt Lake City, Utah	...	...	...

a. Central-reserve city. Brooklyn, which attained reserve status after 1887, is not shown separately be it was consolidated with New York in 1898.

b. Central-reserve city after 1887.

c. Minneapolis and adjacent St. Paul separately recognized as reserve cities, but not shown sepa above.

d. Kansas City, Missouri, recognized as reserve city from 1887 on; Kansas City, Kansas, recogniz reserve city in 1910, but not shown separately above.

largest center, opted for central-reserve status although both were important banking centers in the late 1880s. The anticipated increase in deposits was not thought to counterbalance the privilege of carrying half the required reserves as interest-bearing deposits in New York (Rodkey, 1934: 389).

St. Louis, the fifth-ranking center, was the only city other than Chicago to choose central-reserve status after the enabling legislation. Such status was not warranted by the city's financial importance, however, according to Robertson (1954: 88–92, quoted in Duncan et al., 1960: 311–312). St. Louis banks did attract far more deposits than they had previously. Although part of the increase is accounted for by regional growth, the city's banks did capture deposits from the west and southwest that had gone to New York in the past. Nonetheless, the St. Louis banks never were able to establish the financial strength, resources, flexibility, and stability that might be expected given central-reserve status. After the establishment of the Federal Reserve System which undermined many advantages of central-reserve status without eliminating some of the disadvantages, St. Louis was to return to regular reserve-city status.

Between 1900 and 1910 another nineteen places became reserve cities. No new reserve city appeared east of the Mississippi. In some rapidly growing trans-Mississippi regions which were still being organized economically, competitive complexes developed as cities attempted to establish their hegemony, however. In the Pacific northwest, for example, Portland now faced competition from three Washington reserve cities, Seattle, Spokane, and Tacoma. An even greater free-for-all developed among the Texas cities. Houston first had become a reserve city, a status achieved in 1896. By 1910 Dallas, Fort Worth, San Antonio, Galveston, and Waco also had gained such status. In addition, two cities in Oklahoma joined the competition: Oklahoma City; and Muskogee. Another cluster of rivals that had emerged by 1910 included four Iowa cities: Cedar Rapids and Dubuque in the east; Sioux City in the west; and centrally located Des Moines, which had achieved reserve status in 1890. In Kansas both Wichita and Topeka emerged as reserve cities; Omaha was challenged by neighboring South Omaha; and Kansas City, Missouri had as a competitor its namesake in Kansas. Further west, both Salt Lake City and Pueblo became reserve cities.

The rivalries among western cities did not take the form of a

David battling a Goliath. The larger reserve-cities among the competitors typically were small by the standards of eastern cities. New
York then had five million residents; the populations of nine other
centers exceeded a half million. Spokane and Tacoma, with populations of about 100,000, were pitted against Seattle with its population of a quarter-million, however. The largest of the southwestern
rivals had but a hundred thousand residents. Although east of the
Mississippi were only two reserve cities that did not rank among
the nation's "top thirty" with respect to population size, there were
west of the Mississippi twenty-three reserve cities that did not rank
among the nation's "top thirty." Many small western reserve cities
were of insufficient financial importance to justify their special status, and the status expressed metropolitan aspirations rather than
accomplishments.

Reserve status had, then, a somewhat different significance in
the long-settled parts of the nation than in the more recently opened
territory. In the territory that had been organized before the National Bank Act became effective, designation as a redemption center
was recognition of success on the part of local bankers in developing
a center-oriented commercial network. Outside that territory, the
designation could only reflect the ambitions of local banking communities. There were, to be sure, miscalculations; and some reserve
cities failed to profit by their special position as recipients of country-bank deposits (Rodkey, 1934: 389–391).

(Of incidental interest is the first number appearing above the
fraction sign in the upper right corner of all checks and followed by
a dash, called a "prefix number." Each reserve city existing in 1910
was given a unique prefix number, based on its population rank;
and all other communities in a state were given a residual prefix
number (Rand McNally Bankers Directory, 1951, January ed.; 61).
Even today, Muskogee bank checks have their own prefix number
although Muskogee has fewer than 50,000 residents, but Tulsa
banks share with other places in Oklahoma a state "residual" prefix
although the city has more than 250,000 residents. The only exceptions in the assignment of prefix numbers were: two nonreserve
cities, Buffalo and Memphis, were given their own prefix numbers
because of their size and importance as banking centers; and three
reserve cities failed to receive unique prefix numbers—Brooklyn,
combined with New York City; Kansas City, Kansas, combined with

Kansas City, Missouri; and South Omaha, combined with Omaha.)

Reserve-status fails to discriminate among the major centers, that is, New Orleans and all larger communities, with respect to financial importance. Each, save Buffalo, had elected reserve-city status before 1900. The data are illuminating on two points, however. First, although it was not as traditional metropolises that Pittsburgh, Cleveland, Detroit, Milwaukee, and Washington gained national prominence in the late 1800s, the financial function in each had been sufficiently strong by the 1860s to warrant their inclusion among the initial redemption centers. Second, Minneapolis was not among the first cities on the mid-continent plain to opt for reserve status although, before the turn of the century, it had become established as the region's foremost metropolis.

OTHER FINANCIAL INDICATORS

Although all major centers except Buffalo held reserve-city status, they were not equally popular choices among the nation's bankers for the location of a principal correspondent bank. Particularly before the establishment of the Federal Reserve System in 1913, bank correspondents provided the extralocal linkages required by each bank to meet the financial needs of its community. Pritchard's description of the bank-correspondent network as providing "the sinews of these interbank contacts" is apt; the locations of chooser and chosen tend to define spatially the recurring exchanges between bankers in the greater and lesser financial centers (quoted in Lieberson and Schwirian, 1962: 72).

The *Rand McNally Bankers Directory* which has been published yearly throughout this century lists the principal correspondents of all banks in the United States. A sample of some 500 banks was drawn from the 1900 listing. The bank was classified by the geographic division in which it was located, and each major center in which the bank had a principal correspondent was recorded. The financial strength of a center is gauged by the local concentration of banks chosen as correspondents by banks located elsewhere.

In most instances, banks have principal correspondents in more than one city. Nonetheless, the major centers were sharply differentiated as correspondent-bank locations. Nearly four-fifths of the nation's banks had a principal correspondent located in New York. Chicago occupied a clearly subordinate second-place position, its banks serving as correspondents for just over a third of all institutions. Boston, Philadelphia, and St. Louis were reported as the location of a principal correspondent by about a tenth of the banks. Cincinnati and Minneapolis were chosen by 6 percent of the nation's banks. No other major center of 1900 was selected by more than 4 percent. (Kansas City, which was to displace New Orleans in rank with respect to population size in the 1910s, was selected as a correspondent-bank location by 6 percent of the American banks in 1900.)

New York was the single most frequent choice for a correspondent-bank location on the part of banks in every geographic division of the United States except New England, where Boston enjoyed a slight edge (Table 6-2). There was nationwide recognition of New York's key role in the banking network. Across the northern part of the nation from the Appalachians to the Pacific coast, Chicago was a frequent choice; but a Chicago correspondent was not often selected by banks on the Atlantic seaboard or in the southern states. The correspondent ties to Boston, Philadelphia, and Baltimore, to New Orleans, Cincinnati, and St. Louis, and to Minneapolis, and San Francisco were more restricted territorially. Only banks in a proximate division were likely to select a correspondent in one of these centers.

New York was a truly national metropolis in that bankers in all parts of the nation maintained direct ties with the financial community based in the city. The financial network centered on Chicago was multi-regional in scope, but it did not span the national territory. Only in the north-central area proximate to Chicago or the western territory which had been linked to the settled east by transport routes emanating from Chicago were bankers likely to maintain a direct tie to the Chicago financial community. The financial networks centered on the other cities were restricted to territory proximate to the city, thus distinguishing the regional metropolises from the national metropolis.

New York's special position in the national banking system is reflected in the frequency with which the city was chosen as a correspondent bank location by bankers in the other major centers. New York was the most important correspondent location for banks

Table 6-2. Percentage of Banks in Each Geographic Division Selecting as a Principal Correspondent a Bank Located in a Major Center in 1900 (. . .indicates less than 15 percent selecting)

Location of Correspondent	New Eng.	ATLANTIC Mid	ATLANTIC South	EAST CENTRAL North	EAST CENTRAL South	WEST CENTRAL North	WEST CENTRAL South	Moun-tain	Pa-cific
Atlantic Seaboard									
Boston	69	...	...	...	...	...	...	...	...
New York	62	94	85	75	86	68	86	92	95
Philadelphia	...	49	17	...	...	...	...	...	...
Baltimore	...	...	29	...	...	...	...	...	...
Transmontane East									
New Orleans	...	...	...	...	24	...	27	...	...
Cincinnati	...	...	...	16	19	...	...	...	...
St. Louis	...	...	...	...	21	...	61	...	...
Chicago	...	...	...	63	...	49	...	52	34
Mid-Continent									
Minneapolis	...	...	...	...	...	18	...	...	...
Kansas City*	...	...	...	...	...	15	25	...	...
Far West									
San Francisco	...	...	...	...	...	...	...	23	48

* Displaced New Orleans, smallest of the major centers, in rank with respect to population size between 1910 and 1920.

in each major center except Milwaukee where all local banks maintained correspondents in both neighboring Chicago and distant New York. In no major center did fewer than four-fifths of the local banks maintain a correspondent relation with a New York bank; all banks in most major centers did so.

One of the few other financial indicators which illuminate the relation between center and region is the location of the underwriters of municipal bonds. The geographical destination of the funds obtained by state and local governments in the bond market is always clear-cut. Municipals normally were not sold to the ultimate lenders of the capital, but rather to financial intermediaries who supplied the government body with the funds required and resold the bonds in smaller amounts to institutions and wealthy individuals. Data reported by Goldsmith (1958: 262) show that the commercial banks held a much smaller share of State and local government securities in 1900 than at the present time. Mutual savings banks were much more active in this market; but the share of these bonds owned by individuals was greater then than now.

A sample of municipal bond issues for 1900 was drawn from the *Commercial and Financial Chronicle,* and the underwriters associated with each issue were classified in terms of the city in which they were headquartered. A clustering of underwriters is likely to be indicative of both strong financial resources in the center itself and an extended network of other financial intermediaries, institutions, and wealthy individuals with ties to the center.

In 1900 nearly all issues were purchased by a single bank or investment house serving as the financial intermediary. Cincinnati was the single most important center for bond financing, as measured by the number of issues its firms participated in; its houses were underwriters for about 14 percent of all bond issues floated during the 1900 sample period. Chicago was in second place with 10 percent, barely ahead of New York City. Cleveland, Boston, and Philadelphia were somewhat less important as underwriting centers. No more than 1 percent of the issues was participated in by intermediaries located in any other major center.

Only nine of the sixteen major population centers of 1900 were headquarters location for intermediaries underwriting as many as 5 percent of the issues in any division (Table 6-3). Moreover, intermediaries headquartered in Boston, Philadelphia, New Orleans, San Francisco, Minneapolis, and Cleveland were active only in a single geographic division proximate to the center. Cincinnati's hinterland was a little broader, but only New York and Chicago houses had a hinterland of near national scope. New York houses were active the length of the Atlantic seaboard and in the south-central section of the nation; Chicago houses were active in all sections of the country except the territory along the Atlantic.

On balance, New York with a nationwide hinterland ranked first among the nation's metropolises. The multi-regional character of the Chicago hinterland and the overall strength of the city's financial function secured for it second position among the American metropolises. Among the regional metropolises, Boston and Philadelphia, Cincinnati and St. Louis exhibited greater strength than

Table 6-3. Percentage of Municipal Bond Issues from Each Geographic Division Underwritten by Financial Intermediaries Headquartered in a Major Center in 1900 (. . . indicates less than 5 percent underwritten)

Location of Underwriter	New Eng.	ATLANTIC Mid	ATLANTIC South	EAST CENTRAL North	EAST CENTRAL South	WEST CENTRAL North	WEST CENTRAL South	Mountain	Pacific
Atlantic Seaboard									
Boston	68	...	...	...	...	...	...	...	...
New York	5	30	9	...	8	...	10	...	...
Philadelphia	...	12	...	...	...	...	...	...	...
Transmontane East									
New Orleans	...	...	...	...	...	...	10	...	...
Cincinnati	...	...	9	36	33	...	...	...	...
Chicago	...	...	...	7	17	26	10	33	10
Cleveland	...	...	...	11	...	...	...	...	...
Mid-Continent									
Minneapolis	...	...	...	...	...	9	...	...	...
Far West									
San Francisco	...	...	...	...	...	...	...	...	10

Baltimore or New Orleans, Minneapolis or San Francisco. Neither the five relatively new centers in the manufacturing belt nor the national capital was a focal point in the metropolitan structure as it had evolved by 1900.

Chapter 7

COINCIDENCE
OF FUNCTIONS

The disjunction in the ranking of centers along the financial and manufacturing dimensions which had appeared as early as 1900 gives rise to difficulty if the nation's cities are conceived as a strictly hierarchic system (see Duncan et al., 1960: ch. 3, for one discussion). New York's position as "first city" in the American system seems indisputable. A weaker case can be made for the designation of Chicago as the "second city." With lessening confidence, Philadelphia and Boston might be assigned the third and fourth ranks. What city, then, is to occupy fifth rank? Pittsburgh and St. Louis are the contenders. Their resident populations were similar in size, more numerous than the residents of any center save the "big four." St. Louis clearly outranked Pittsburgh, in fact was coordinate with Philadelphia and Boston, along the financial dimension. On the other hand, Pittsburgh was clearly stronger than St. Louis, more nearly the equal of Boston, along the manufacturing dimension.

The positioning of St. Louis and Pittsburgh in an urban hierarchy is not, of course, the real issue. At stake is the usefulness of the notion of hierarchy for understanding the system of cities that had evolved in America by the turn of the century. That there existed a system is not in question. Differentiation among centers and, hence, dependence among them had sharpened. Differentiation had, however, taken the form of specialization in function or role in the national economy; and it is this fact that makes conceptualization of the system of cities as a graded order problematic.

In the financial realm or with respect to the traditional metropolitan function, a tendency toward hierarchy was discernible. Among the nation's large population centers, we can identify some places which had no extralocal financial significance. Financial networks of regional scope were focused on several places, and a multi-regional network converged on Chicago. A near-national network, representing both direct linkages from outlying financial institutions and indirect linkages by way of the regional metropolises, was centered on New York and distinguished it as the hub of the national money market. Although an individual place might shift to a higher or lower order were financial significance construed more or less broadly, metropolises seemingly could be assigned to graded orders on the basis of scale and character of local financial activity.

At a time when commerce was the sole city-building activity, we can suppose that the system of cities could be understood as a graded order. Change in the system would have been generated, in part, by the extension of territory tributary to the metropolitan-based organization, a process which both modified the relative positions of established centers near the top of the hierarchy and created new centers at the lower orders of the hierarchy. Simultaneously generating change in the system would have been economic growth differentials among the uncontested hinterlands of the respective established centers and, less often, a redistribution of tributary territory between centers.

The main features of the American urban system as it had developed by, say, 1860 are captured by this imagery of a graded order of commercial centers in which change reflected alteration in tributary territory. This metropolitan system did not disappear in the late 1880s, but it ceased to be identical with the American system of cities. Large-scale manufactures became an alternative to commerce as a city-building activity. The graded order of commercial centers was no longer adequate to account for the roles or sizes of cities in the national economy.

Had manufactures remained an adjunct to the metropolitan function, increasing only in scale of activity, the tendency toward hierarchy in the system of cities would have remained clear-cut. In point of fact, concentrations of manufacturing activity developed in the environs of some, but not all metropolises; moreover, equally large concentrations of manufacturing activity developed in the environs of some lesser places which had no extralocal financial significance. A system of centers, specialized in function, falling into no single graded order, had emerged.

The fortunes of these new manufacturing centers did not rest on their success in organizing production in hinterland territory and markets for hinterland products, thereby ensuring flows of commodities and money through local commercial institutions. In fact, the prosperity of many local manufacturers had no distinctly "regional" base. Their suppliers and customers were scattered through the national territory or, on occasion, clustered within the immediate environs of the manufacturing center itself. The fortunes of the manufacturing centers depended on an ability to retain and attract lines of manufacture in which production was expanding in response to a growing "national" demand.

An hierarchy of manufacturing centers is difficult to conceive, save as simply an order based on scale of manufacturing activity and, possibly, diversity of manufacturing specialties. The spatial organization of the production of goods does not yield to the same sort of conceptualization as the spatial organization of the provision of services; and the manufacturing center had a "product" to sell, in contrast to the metropolis which sold a "service." New York financial institutions, for example, offered for sale the full range of services offered by, say, Cincinnati institutions; but the New York institutions offered, in addition, some special services which were used by both New York and Cincinnati customers. Philadelphia manufacturers produced a range of goods somewhat broader than, but distinct from, say, Pittsburgh manufacturers; thus, the "direction" of dependence cannot be established readily.

The system of centers that had evolved by 1900 can be more adequately represented as the outcome of two imperfectly related hierarchic systems than as a single graded order. The order of centers with respect to importance in the financial network, the key indicator of the metropolitan function, is fairly unambiguous. At the apex is New York; in second position is Chicago, focus of a multi-regional network; forming a third order are Boston, Philadelphia, Cincinnati, St. Louis, New Orleans, San Francisco, and Minneapolis, each the focus of a regional network; with little more than local financial significance are Baltimore, Pittsburgh, Buffalo, Cleveland, Detroit, and Milwaukee. The introduction of finer distinctions might recognize Boston and Philadelphia as the more important third-order centers, Baltimore and Pittsburgh as the more important fourth-order centers. In the manufacturing hierarchy as in the metropolitan hierarchy, the foremost centers are New York and Chicago, respectively. Among the third-order metropolitan centers, however, Boston, Philadelphia, Cincinnati, and St. Louis are distinctly more important as centers of manufacture than are New Orleans, San Francisco, and Minneapolis. Moreover, the fourth-order metropolitan centers of Baltimore and Pittsburgh are coordinate with the third-order metropolitan centers of Boston, Philadelphia, Cincinnati, and St. Louis along the manufacturing dimension.

The special-function city of Washington is not easily accommodated in the system represented as the outcome of two imperfectly related hierarchic systems, one with respect to distinctively metropolitan pursuits, the other with respect to manufacture. In either system, Washington would occupy an order far lower than that occupied by any other major center. Nonetheless, Washington matched the smaller of the major centers in size of its resident population.

We are inclined to argue that the role occupied by Washington was itself a creation of the disjunction in the ranking of centers along the metropolitan and manufacturing dimensions. The outstanding feature of the role was regulation, not voluntary, but rather officially sanctioned. Disjunction in ranking and the differential bases of change in the two systems simultaneously intensified intercity competition and weakened the metropolitan structure as a regulatory mechanism. Spatially, the economic ties of large-scale manufacturers crosscut the nested regions and grade-ordered metropolitan centers through which commercial activity had been organized.

Perhaps it is most fruitful to conceive the system of major centers simply as an assemblage of functionally differentiated and, thereby, interdependent communities. The number of distinguishable roles in the system is as great as the number of centers in the system, but there is sufficient similarity in the roles of, say, New Orleans and San Francisco or Pittsburgh and Cleveland with respect to economic base and regional orientation to justify a set of broadly defined roles.

To be found in the system are: traditional metropolises in which manufactures were an outgrowth of the community's regional commercial primacy; metropolises in which diversified industrial manufactures had located; centers of industrial manufactures wherein distinctively metropolitan pursuits were overshadowed by manufacturing might; and, of course, the "control" center of the system.

In the American setting, the role of traditional metropolis antedated the role of industrial-manufacturing center. The historical record does not support the notion of an inevitable developmental sequence whereby the metropolis is transformed into a center of industrial manufactures, however. On the contrary, the spatial organization of "services" appears to have evolved more or less independently of the organization of "production." Nodes in the service and production networks could, but need not coincide in space.

PART II

NEW CENTERS
IN THE
ESTABLISHED SYSTEM

THE ELECTRIC-OIL-AUTO COMPLEX

Chapter 8

Any history of American ideas about American metropolitanism would refer to the contributions of Gras in the 1920s and McKenzie in the 1930s. A brief history would identify with McKenzie the idea of the metropolitan community as a type of settlement pattern; Gras's name would be linked with the idea that a particular array of functions distinguishes the metropolis from other communities (Duncan et al., 1960: 83–85). Their respective emphases, complementary rather than conflicting, are perhaps best understood in relation to events of the decade which separates their contributions.

Gras (1922: 186, for example) argued that the metropolis is distinguished by its role in coordinating long-distance or intermetropolitan trade although it also is a focus of local or intrametropolitan trade and may include large-scale manufactures in its array of functions. During the 1920s, the railroads began to lose ground to motor transportation. When railways had become an alternative carrier to waterways some decades earlier, the intersection of rails came to confer a long-distance transport advantage as great as that of a waterways junction. McKenzie called attention to the fact that the motor vehicle was not altering the long-distance transport advantage of one community vis-à-vis another, but rather modifying the settlement pattern in the environs of major population centers.

119

McKenzie as well as Ogburn, another perceptive social analyst of the 1930s, emphasized the influence of the motor vehicle on intra-metropolitan as opposed to intermetropolitan relations. McKenzie (1933: 139–140) wrote:

> . . . as rail transportation gradually surmounted the physical barriers that formerly separated these different areas [oriented to various water systems] and brought erstwhile remote population groups into direct economic relations with each other, not merely in the mutual exchange of products but in competition for common markets, cities rather than geographic sections became the nuclei in the larger economy. . . . Coming after the main outlines of settlement pattern were well established, the motor vehicle and the motor highway have not materially changed this basic structure. They have created no new large cities. They have, however, effected modifications in local relations which, in the aggregate, are perhaps quite as significant as those introduced by the railroads.

Ogburn (1937: 4) seemed to concur, noting that:

> Just as the railroads caused cities to spring up all over the country, so the automobile is changing them, hurling their population with a centrifugal force outward into the suburbs and drawing into an ever-widening trading area many millions of inhabitants drawn from remoter regions. Thus it has created a new unit of population neither city, town, nor hamlet, for which there is as yet no name, but which is often referred to as a metropolitan area.

The emphasis is proper, for the motor vehicle remains primarily a short-haul carrier extending the territorial limits of the "local community" identified with the center through recurrent use of center-based commercial and service facilities.

The assertion that new large cities were created by the railroads, but not by motor transportation is not wholly tenable if it is taken literally. New Orleans, smallest of the centers which had occupied the top five ranks in the city-size distribution from 1820 to 1860, was displaced in rank by eleven centers between 1860 and 1900; in contrast, only two centers shifted ahead of New Orleans between 1900 and 1940. The frequency with which new centers broke into the ranks of the established centers slackened, however, before motor transportation can have had any impact on the territorial organization of economic activity. Moreover, in very few instances can the case be made that railroad transportation as such created the new center.

Railroads must be taken as a symbol of the steam-and-steel complex which transformed American manufactures and created new centers of industrial manufactures in settled parts of the nation at the same time that trade generated in newly opened trans-Mississippi resource regions was creating new regional metropolises. Motor transportation was imbedded in the electric-oil-auto complex which became nationally prominent after the manufacturing-belt had been demarcated and the national territory had been largely integrated into the metropolitan-based economic organization.

Aspects of the electric-oil-auto complex other than motor transport seem equally unlikely to have altered the relations among established centers or to have created new large centers. The need for concentrated industrial activity lessened as relatively more machines were powered by electric motors driven by purchased energy generated outside the establishment, for example. Until late in the nineteenth century, power had been transmitted from prime movers to power-driven machines "entirely by mechanical means such as belts, chains, ropes, discs, cams, levers, gearing, and shafting"; then, "electricity proved to be a practical link for transmitting power over great distances" (Potter and Samuels, 1937: 254). Again, the pipeline was a long-distance, but highly specialized carrier. As Wilson (1942: 469) observed, "most water cargoes are originated inland and move by pipe to ports for transshipment." The potential of oil, either its extraction or shipment, for "city building" was slight.

GROWTH OF THE COMPLEX

Although census statisticians seldom are given to hyperbole in their official reports, the special report on electrical apparatus and supplies in 1900 opened with reference to a remark about telephony made in the census report of 1880: "At the beginning of 1879–80 this business amounted to little or nothing; at the end of the year it represented one of the great interests of the country." The special report continues with:

Not only is this again singularly true of the present condition of telephony, twenty years later, as to its sudden expansion, but it applies forcibly along the whole range of electrical industries and applications . . . in 1900 the unprecedented adoption of the electric motor for power transmission, factories, etc., as well as for the automobile, would have offered further proof of the rapidity of movement. . . . [Twelfth Census, Vol. 10, Pt. 4: 153]

The production of electrical machinery, apparatus, and supplies was concentrated in the northeastern manufacturing belt and had become an industrial specialty in the established centers of Boston, New York, Pittsburgh, Cleveland, and Chicago. Four-fifths of the work force in the industry could be found in six states of the northeast as late as 1920, with the individual state shares ranging from 18 percent for New York to 11 percent for Pennsylvania and Massachusetts, respectively. (Statistics are drawn from Fourteenth Census, Vol. 10.)

Although the automobile is cited in connection with "adoption of the electric motor" in the 1890s, motor vehicles were simply subsumed under the carriages-and-wagons title in the detailed industrial classification of 1900. Subsequent retabulation has shown that there were then 2,500 workers employed in the industry. About 4,000 vehicles were manufactured; four-tenths were electric-powered; and Massachusetts out-paced all other states in production. Five years later, the industry claimed 13,000 workers; and production had become concentrated in Michigan where 3,000 workers manufactured 9,000 vehicles, all gasoline-propelled.

Closely identified with the development of vehicles propelled by gasoline engines in America was a Michigan resident, Henry Ford, who had produced a vehicle driven by a water-cooled engine as early as 1896, founded the Ford Motor Company in 1903, and by 1908 was mass-producing the famous Model "T" (Derry and Williams, 1961: 395, 607). By 1910 fewer than four percent of the vehicles produced were electric-powered, and Michigan had consolidated its position as the center of the new industry. The state's establishments reported a work force of 28,000 out of a national total of 85,000 and produced 65,000 of the 127,000 vehicles manufactured. A decade later, Michigan still ranked first among the states with respect to share of the industry: 142,000 of the nation's 243,000 workers in the "automobile" industry; and 55,000 of the 153,000 workers in the "automobile bodies and parts" industry. The Michigan "motor city" of Detroit, in fact, had more workers in the industry than did the second-ranking state, Ohio. (Statistics are drawn from Twelfth Census, Vol. 10, Pt. 4, Special Reports on "Carriages and Wagons" and "Electrical Apparatus and Supplies," and Thirteenth Census, Vol. 8: 472 and 553; and Vol. 9: 565; and Fourteenth Census, Vols. 9 and 10.)

Ultimately the automobile generated a heavy demand for products of the petroleum-refining industry, but illuminating oils were the industry's most valuable product in 1900. Nearly a fourth of the nation's crude petroleum was being produced in the state of Pennsylvania, and slightly over a fourth of the petroleum products came from Pennsylvania refineries. From producing areas in Pennsylvania and adjoining states, crude oil was moving by pipeline to "large refineries on the seaboard, near New York, Philadelphia, and Baltimore, and on the shores of Lake Erie, near Buffalo and Cleveland" (Twelfth Census, Vol. 8, Pt. 2: 752). About two-fifths of the national output was destined for overseas markets, where illuminating oils were in heavy demand (Twelfth Census, Vol. 10, Pt. 4: 686).

At least one refinery was reported to be in operation in each of four states west of the Mississippi. The descriptive text accompanying the statistical tables for 1900 made no reference to the petroleum industry in Colorado or Texas, however, noting rather the local availability of coal reserves. The report for Kansas observed that "More than a hundred wells, producing either oil or gas, have been opened during the last few years" in the southern part of the state; found in the same locality as coal and zinc ore, the new fuel source was stimulating expansion in the smelting and refining of zinc (Twelfth Census, Vol. 8, Pt. 2: 259). The new fuel source received even more attention in the California report:

> Coal is found in the state, but it is poor in quality. A large part of the coal consumed is imported. . . . It has recently been discovered that California possesses abundant stores of petroleum, and many wells have been sunk, especially in the southern part of the state. The importance of this discovery to the manufacturers of California can scarcely be exaggerated, since it practically solves the problem of fuel. Its influence is only slightly shown by the returns of the Twelfth Census, for the exploitation of the oil fields was but

just beginning during the census year. [Twelfth Census, Vol. 8, Pt. 2: 33–34]

A little less than a tenth of the crude petroleum consumed by American refineries in 1920 was extracted in the Appalachian field (Pennsylvania grade). Half originated in the American mid-continent field; a fifth came from California; and a tenth was of Mexican origin. Three-tenths of the refining, however, was still carried out in Pennsylvania and New Jersey although the Appalachian field was no longer the major supply area for crude oil. Only three-tenths of the refining occurred in the mid-continent states of Kansas, Missouri, Oklahoma, and Texas near the major supply area; and only a tenth took place in California. (Statistics are drawn from Fourteenth Census, Vol. 10: 758 and 762.) From the mid-continent producing fields, crude oil was moving through pipelines to refineries on the mid-Atlantic coast; and these flows were supplemented by movements of water carriers from Pacific and Gulf ports to Atlantic ports (Wilson, 1942: 459).

At least two bases of justification for selecting the electric-oil-auto complex as the symbol of the American economy in the early 1900s can be suggested. First, their respective shares of all capital in the manufacturing sector increased more rapidly than the share of any other branch of industry in the three decades following 1900 (Figure 8-1). Increases were recorded also in the shares of the metal industries, both iron and steel and the nonferrous group, and in the paper, chemicals, and tobacco branches of manufacture; but they did not match the gains in petroleum refining, or in the motor-vehicle or electrical-machinery industry.

An alternative basis for singling out electricity, motor vehicles, and oil might be their growing importance with respect to power in the manufacturing process, the movement of people and goods, and fuel supply, respectively (Figure 8-2). A series on use of motor vehicles, expressed as vehicle-miles traveled, dates back to 1921 and exhibits a strong upward secular trend over the subsequent two decades; no corresponding upward secular trend appears in the series on use of rail transportation, either with respect to passenger travel or freight transport, however. The production of crude oil increased substantially in the early 1900s, but the trend line associated with the amount of bituminous coal mined is flat. The final

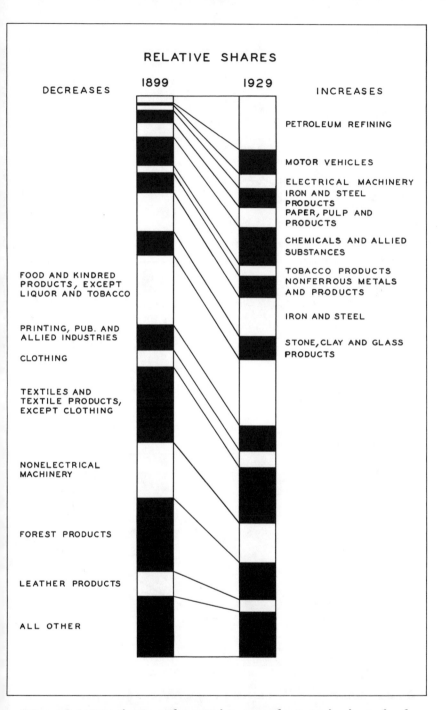

Figure 8-1. Distribution of capital in manufactures by branch of industry: United States, 1899 and 1929. (Source: *Historical Statistics of the United States,* Series P82–133.)

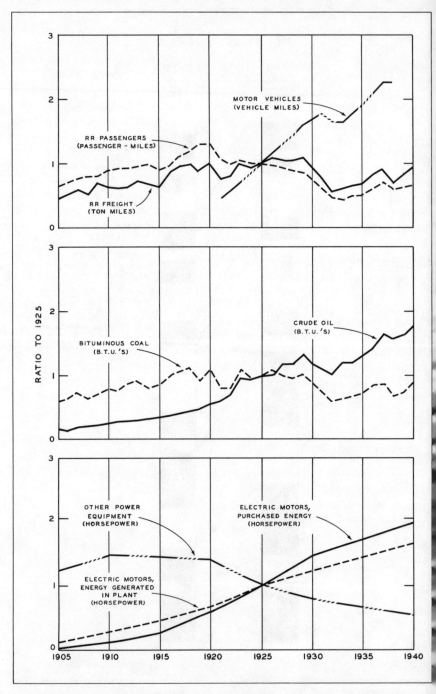

Figure 8-2. Selected indicators of transport modes, mineral energy fuels production, and power equipment in manufacturing establishments: United States, 1905 to 1940. (Source: *Historical Statistics of the United States,* Series M72–74 and Q67, 82, 321; Sixteenth Census, Manufactures, Vol. 1, p. 275.)

series pertain to the horsepower available in manufacturing establishments and show that the use of electric motors as the link between prime movers and machinery to be powered became steadily more pervasive.

INFLUENCE ON CITIES

Not until the 1940s when Houston, a port proximate to the mid-continent oil fields, penetrated the ranks of the established centers can a direct link between the electric-oil-auto complex and the rise of a new center be established. Four decades earlier, however, Los Angeles had established itself as a major center; and its newfound prominence traced indirectly to the complex.

The southern California community of Los Angeles had only 10,000 residents in 1880. San Francisco, the metropolis of the far west, then had more than 200,000 residents. By 1910 Los Angeles residents numbered nearly 400,000, making it the fifteenth largest center in the nation; San Francisco with 600,000 residents ranked seventh. A decade later, Los Angeles had 700,000 residents and had shifted upward into tenth position; San Francisco with 800,000 residents had slipped into ninth position.

Los Angeles showed strength neither in the manufacturing sector nor in the financial sphere in 1900 by comparison with the established centers. The 1900 benchmark measurements were taken at the beginning of the decade during which Los Angeles penetrated the ranks of the established centers, however. It was also at this time that exploitation of the proximate oil fields was gaining momentum. For possible clues about the bases of the center's new prominence, the 1920 industrial structure of Los Angeles is contrasted with the 1920 structures of New Orleans, a traditional regional metropolis, and Cincinnati, a regional metropolis and center of industrial manufactures (Table 8-1).

The Los Angeles profile was weighted heavily with extractive activity. Workers in the trade and service sector, notably retailing, insurance and real estate, and professional services, also were over-represented. These distinctive features of the Los Angeles structure may simply reflect the recency of growth in the center. Alternatively, they may suggest that given a sufficiently extended "metro-

Table 8-1. Percentage Distribution of the Work Force by Type of Activity in 1920 and Percentage-Point Change, 1910–20, for New Orleans, Cincinnati, Los Angeles, and Kansas City

Type of Activity	Distribution				Change			
	NO	Ci	LA	KC	NO	Ci	LA	KC
Trade and Service	52.6	46.4	55.2	54.2	−2.8	−0.2	0.1	1.1
Retail trade[a]	10.6	10.6	13.2	13.0	−2.4	−1.6	−1.5	−0.7
Finance[b]	0.6	0.5	0.8	1.0	0.0	0.1	0.0	0.3
Insurance and real estate[c]	0.8	0.8	2.2	1.4	0.1	0.2	−0.4	−0.3
Other trade	2.5	2.3	2.4	3.1	−0.5	0.3	−0.3	0.5
Entertainment[d]	0.7	0.8	3.1	1.0	−0.1	−0.3	1.4	−0.1
Other professional service	4.4	5.0	7.9	5.3	0.4	1.0	0.9	0.2
Clerical occupations	11.8	12.4	11.4	14.9	3.4	3.1	2.1	3.9
Public service	3.5	1.8	2.0	2.0	1.2	0.1	0.5	0.2
Domestic and personal service	17.7	12.2	12.2	12.5	−4.9	−3.1	−2.6	−2.9
Manufacturing and Mechanical Industries	32.0	43.8	31.6	33.0	1.9	−0.2	0.6	−0.6
Outside manufacturing establishments	17.0	6.2	13.5	10.6	−1.6	−2.9	−5.7	−5.6
In manufacturing establishments[e]	15.0	37.6	18.1	22.4	3.5	2.7	6.3	5.0
Slaughtering and meat packing	0.0	0.9	0.5	8.0	0.0	0.3	0.2	2.0
Foundry and machine-shop products	0.7	7.6	1.9	1.5	0.2	2.6	0.3	0.7
Printing and publishing	0.5	2.3	1.1	1.3	−0.2	0.1	0.1	−0.2

	(1)	(2)	(3)	(4)	(5)	(6)	(7)	(8)
Bread and other bakery products	0.9	0.9	0.4	1.3	0.3	0.1	−0.4	0.4
Canning and preserving	0.1	0.0	0.9	⎫	0.0	−0.1	0.6	⎫
Women's clothing	0.1	0.5	0.7	⎬ 10.3	0.1	−0.3	0.6	⎬ 2.1
Other industries	12.7	25.4	12.6	⎭	3.1	0.0	4.9	⎭
Transportation and Communication	14.2	8.8	8.6	12.0	2.2	0.3	−1.4	−0.1
Telephone and telegraph	1.0	1.1	1.3	1.6	0.4	0.3	0.2	0.5
Water transportation	5.0	0.1	0.4	0.0	⎫			⎫
Steam railroad	3.0	2.6	2.3	5.1	⎬ 1.8	0.0	−1.6	⎬ −0.6
Other	5.2	5.0	4.6	5.3	⎭			⎭
Resource Extraction	1.2	0.9	4.6	1.0	−1.3	0.0	0.6	−0.3
Agriculture and forestry	1.1	0.8	3.8	0.8	−1.4	0.0	1.1	0.0
Minerals	0.1	0.1	0.8	0.2	0.1	0.0	−0.5	−0.3

a. Clerks, floorwalkers and foremen, laborers, porters and helpers, salesmen and saleswomen, in stores; decorators, drapers, and window dressers; newsboys; retail dealers; undertakers.

b. Bankers, brokers, and money lenders; officials of insurance companies.

c. Insurance agents; real estate agents and officials.

d. Actors and showmen; musicians and teachers of music; photographers; teachers (athletics, dancing, etc); theatrical owners, managers, and officials.

e. Wage earners as reported in Census of Manufactures.

Note: Full identification of caption symbols appears in Table 5-1.

Source: Fourteenth Census, Vol. 4, Table 19 and Vol. 9, State Table 30, 31, 36, or 46; Thirteenth Census, Vol. 4, Table 3 or 4 and Vol. 9, State Table 3.

politan community," so-called local trade and service activities can
support a sizable core and that the core itself can include a diversity
of land uses.

Recent changes in the industrial structures of the three centers
call attention to the growth of the manufacturing sector in Los
Angeles. The proportion of the work force engaged in manufactures
increased substantially more in Los Angeles than in the older
centers during the 1910s, and no line of industry can be singled out
as the major source of increase. It is tempting to argue that the
discovery of oil reserves in southern California shifted industrial
development in the far west away from San Francisco, the metropolis through which the region's extractive activity had first been
organized. The remoteness of the region from the eastern manufacturing belt fostered a diversified regional manufacturing base,
reflected in the Los Angeles structure by an absence of distinctive
features.

To say that the electric-auto-oil complex caused the growth of
Los Angeles would be inaccurate, but it provided a relatively favorable setting for the growth of industry in southern California.
Manufacturers in Los Angeles relied far more heavily on oil for fuel,
for example, than did manufacturers in the old centers, and a larger
share of the primary horsepower used in the manufacturing process
was provided by electric motors operated by rented current in Los
Angeles than in either New Orleans or Cincinnati (Fourteenth
Census, Vol. 9). Track operated by electric railway companies increased from 2,000 to 3,000 miles between 1907 and 1917 in California, as compared with increases from 3,700 to 4,200 miles in
Ohio and 200 to 300 miles in Louisiana (*Census of Electrical Industries: 1917, Electric Railways: 26*). As early as 1898 "Los Angeles
County began experimenting with oil" to improve the surface of
roads (Rose, 1950: 87), and by 1904 "California had a substantial
mileage of oiled earth roads . . . and had actually built eighteen
miles of petrolithic roads" (Miller, 1950: 101). Vehicle-registration
series of the Bureau of Public Roads make clear the pace-setting
role of California in the diffusion of the auto after the turn of the
century.

Although Kansas City broke into the ranks of the established
centers a decade later than Los Angeles, it had had regional, if not
national, prominence much longer. When Los Angeles had been a

town of 10,000, Kansas City had had nearly 60,000 residents. As early as 1900, meat packing had appeared as a local industrial specialty; and Kansas City had centered upon it a financial network of regional scope. Salient in the center's 1920 industrial profile were the financial and wholesaling activities characteristic of a regional metropolis, the slaughtering and meat-packing specialties which had a clear tie to extractive activity in the surrounding territory, and railroading by means of which Kansas City tapped the mid-continent tributary territory.

No center penetrated the ranks of the established centers, that is, displaced in rank New Orleans, during the 1920s or the 1930s. There were occurring, however, some noteworthy shifts in rank among the established centers themselves. The 1910s and 1920s had been decades during which Detroit enjoyed a pronounced growth advantage vis-à-vis other major centers. A growth differential favoring Washington became discernible in the 1930s. (Growth differentials among the established centers of 1900 are shown in Figure 8-3.)

The relatively rapid growth of Detroit can be traced to the localization of motor-vehicle production in the center. That it should have been Detroit, rather than another place in the manufacturing belt, that became the "Motor City" is perhaps fortuitous. That the motor city should have emerged from among the places in the manufacturing belt with a nucleus of metal-working specialties and ready access to suppliers and buyers is less a matter of chance. Growth in Detroit slackened in the 1930s with the onset of the great depression.

"Bad times" in the nation were, in a sense, "good times" for the nation's capital where the federal government was a foremost employer. Federal civilian employment in Washington fell sharply from a World War I peak to a 1926 low. In contrast, the manufacturing work force remained stable; and employees in the finance, insurance, and real-estate industries became more numerous. By the mid-1930s, sharp annual increases were being recorded in the number of Washington-based civilian employees of the federal government as federal activity expanded in response to the economic slump. The rapid growth of federal civilian employment was to be sustained by mobilization for World War II. (The timing of increases in federal civilian employees in Washington is contrasted

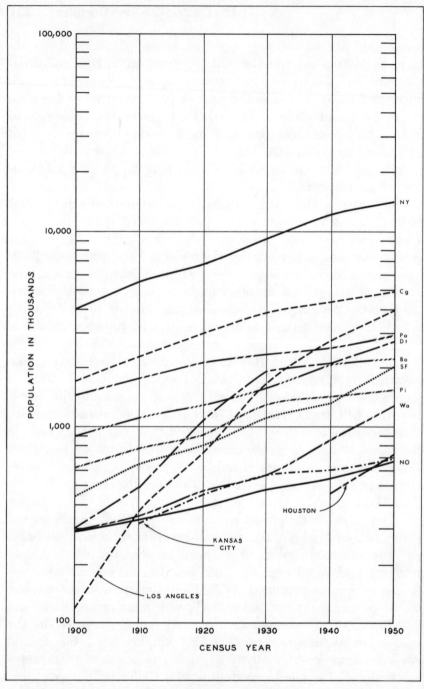

Figure 8-3. Populations of cities greater than or equal to New Orleans in population size: 1900 to 1950. (Seven centers—St. Louis, Cleveland, Baltimore, Minneapolis, Buffalo, Milwaukee, Cincinnati—with populations greater than New Orleans and less than Pittsburgh at each date not shown.)

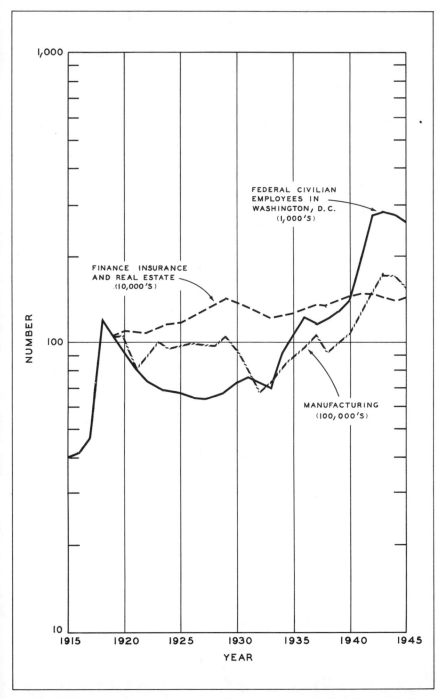

Figure 8-4. Employees in selected activities: United States, 1915 to 1945. (Source: *Historical Statistics of the United States,* Series D51, D54, and Y242.)

with the timing of change in the nation's manufacturing work force and employees in the finance, insurance, and real-estate industries in Figure 8-4.)

From many perspectives, the American scene was transformed as markedly in the first four decades of the twentieth century as in the last four decades of the nineteenth century. From the perspective of the relation between metropolises and resource regions or from the perspective of the division of labor among major centers, however, the salient feature of the twentieth century was the stability of the system.

Chapter 9

SINCE 1940

A 1900 bench mark was established for the manufacturing and financial functions of the leading centers, but not until 1940 can a bench mark be established with respect to the full range of economic activity. A reading on local economic activity in any given year may be misleading because of some unusual circumstance, and measures on manufacturing, wholesaling, and service activity at the end of the 1930s may be particularly suspect if areal differences in the impact of the great depression are assumed. Fortunately, however, measures can be obtained for 1947–1948, 1954, 1958, and 1963 (from a Census of Manufactures or Business conducted in the given year) which are comparable with the initial readings of 1939 (based on data of the Sixteenth Census). Insofar as intercenter differences persist or the patterns of change are similar over variable time spans, we can be reasonably sure that we are not misreading the basic structure and the way in which it is modified.

Up to this point, we have been unable to contrast the economic structures of the leading centers with the structures of lesser communities. Extreme functional specialization is, of course, characteristic of the small community, not the great city. Such popular designations as the "Iron City" or the "Motor City" notwithstanding, it seems clear that no single industrial specialty could support a

million-odd local residents. Nonetheless, the functional differentiation
among the largest centers is more easily evaluated when differentiation among a broader group of communities is the norm. Accordingly, we have compiled 1940 benchmark measures for eighty-
four sizable metropolitan areas. Included among them are the
eighteen largest centers of 1940, Houston which displaced New
Orleans in rank during the 1940s, and three places which would
break into the ranks of the established centers in the 1950s—Dallas,
Seattle, and Miami.

THE MANUFACTURING-COMMERCE MIX

A coincidence of strong wholesaling and manufacturing functions distinguished the leading centers from lesser places although
the "mix" varied widely among the leading centers themselves. The
major centers had relatively high levels of manufacturing activity
as compared with lesser places in which the level of wholesaling
was similar. Among communities with similar levels of manufacturing, the major centers had a relatively high level of wholesaling
activity (Figure 9-1).

Los Angeles, the most rapidly growing of the major centers,
occupied a less favorable position with respect to levels of wholesaling and manufacturing than any other established center except
the national capital of Washington. The per capita value of manufactures in Los Angeles was lower than in any major centers save
New Orleans and Washington. With respect to the per capita value
of wholesale sales, only Washington and the manufacturing-belt
centers of Baltimore, Philadelphia, Pittsburgh, Buffalo, and Milwaukee ranked lower than Los Angeles. Among the centers which were
to gain national prominence in less than twenty years, only Miami
evidenced both wholesaling and manufacturing functions weaker
than those observed in Los Angeles. Houston, Dallas, and Seattle
were relatively strong in the wholesaling sector although the local
manufacturing base was weak.

During the 1940s, there was strong persistence in the relative
positions of centers with respect to per capita levels of manufacturing activity and wholesaling, as well as to the per capita volume of

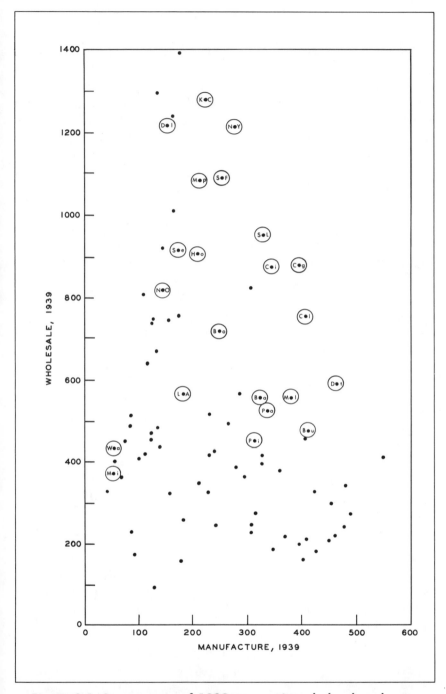

Figure 9-1. Scattergram of 1939 per capita wholesale sales on 1939 per capita value added by manufacture for metropolitan areas (84 areas with 1960 populations of 250,000 or more and measures of sales and value available at four dates subsequent to 1939).

activity in a third sector of the economy indexed by service receipts. When per capita value added by manufacture in 1947 is regressed on per capita value added by manufacture in 1939 over the eighty-four sizable metropolitan areas, the coefficient of correlation is found to be .95. When per capita wholesale sales in 1948 is regressed on per capita wholesale sales in 1939 over the eighty-four places, the coefficient of correlation is found to be .96. The corresponding interannual correlation with respect to per capita service receipts over the seventy-nine places for which a complete time series can be compiled is .84.

Again in 1947–1948, Los Angeles is found to have relatively low per capita levels of wholesaling and manufacturing activity although with respect to per capita service receipts Los Angeles was outranked only by New York, Chicago, and San Francisco among the major centers (Figure 9-2). The statistical evidence on change in levels of activity does not, however, suggest that the high per capita service receipts were growth induced or that rapid population growth depressed the per capita levels of wholesaling or manufacturing. When the 1947 or 1948 per capita measure is regressed on both the corresponding 1939 per capita measure and a measure of recent growth (namely, the 1939 population as a percentage of the 1947 or 1948 population), the net regression coefficients in standard form are found to be:

	1939 per capita	Recent growth
Manufacturing	.95	.01
Wholesaling	.99	.12
Services	.85	.03

There is, in fact, a rather remarkable long-run stability in the relative positions of communities with respect to level of activity in each of the three economic sectors. Between successive observations in the time series on per capita level of manufacturing activity, the regression coefficients in standard form (or correlation coefficients) are found to be:

| 1947 on 1939 | .946 | 1958 on 1954 | .962 |
| 1954 on 1947 | .944 | 1963 on 1958 | .956 |

On the assumption of a simple chain model, a set of six interannual correlation coefficients over spans ranging from nine to twenty-four

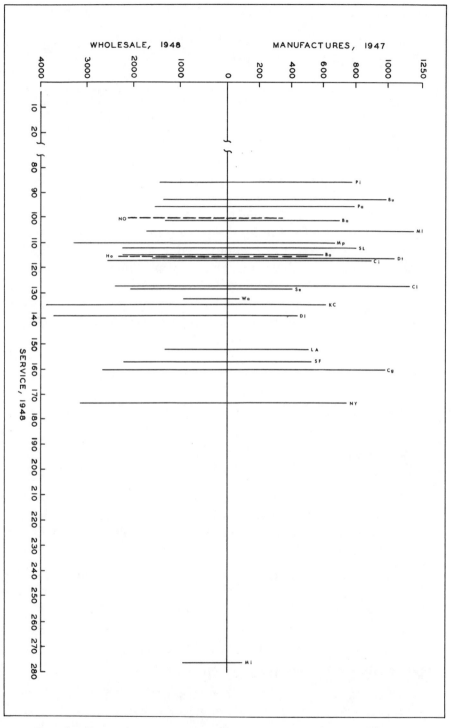

Figure 9-2. Per capita service receipts in relation to per capita wholesale sales and per capita value added by manufacture, 1947–48, for major metropolitan centers.

years can be calculated and compared with the corresponding set of observed correlation coefficients:

	Calculated	Observed		Calculated	Observed
1939–1954	.893	.888	1948–1958	.908	.914
1939–1958	.859	.855	1947–1963	.868	.900
1939–1963	.821	.866	1954–1963	.919	.915

Although some modest departures of observed from calculated values can be noted, the "fit" is reasonably close. This can be taken to mean that the pattern of intercenter differences is disturbed over time by small, though cumulative "random shocks." Projected indefinitely far into the future, the pattern of intercenter differences would no longer resemble that observed at mid-century; but assuming interannual correlations of about .95 per quinquennium and a simple causal chain as the model of change, a coefficient of correlation as high as .6 will be found between the per capita levels of manufacture in these centers in the year 2000 and the 1950 per capita levels. (On chain model, see Duncan, 1966.)

Stability over time in the relative positions of centers with respect to per capita wholesaling appears even stronger than the persistence in per capita levels of manufacturing. The regression coefficients in standard form between successive observations in the time series on per capita wholesale sales are found to be:

1948 on 1939	.957	1958 on 1954	.971
1954 on 1948	.981	1963 on 1958	.982

Again the simple causal chain seems to provide a more or less adequate model of the pattern of change. The calculated and observed coefficients of correlation over the six possible periods are found to be: ·

	Calculated	Observed		Calculated	Observed
1939–1954	.939	.931	1948–1958	.953	.962
1939–1958	.912	.912	1948–1963	.936	.947
1939–1963	.896	.895	1954–1963	.954	.956

A corresponding set of interannual regressions with respect to per capita service receipts takes on the values:

1948 on 1939	.813	1958 on 1954	.973
1954 on 1948	.896	1963 on 1958	.943

Persistence over the first fifteen years covered by the time series appears slight relative to the stability manifest since 1954. Perhaps not too much substantive significance should be attached to the apparent tightening of the pattern, however, in view of the difficulties in compiling a fully comparable time series. For the same reason, the possibility that the pattern of change is captured by a simple chain model cannot be discounted although the "fit" is poor over time periods which extend further back than 1948. The calculated and observed coefficients are:

	Calculated	Observed		Calculated	Observed
1939–1954	.729	.827	1948–1958	.871	.889
1939–1958	.709	.818	1948–1963	.822	.812
1939–1963	.668	.821	1954–1963	.917	.907

At no observation date does a measure of recent population growth have an effect on per capita value added, net of the effect of per capita value added at the preceding observation date. When per capita value added in, say, 1963 is regressed on per capita value added in 1958 and on the recent growth measure (the ratio of 1958 to 1963 population), the regression coefficient associated with the recent growth measure is found to have a value less than its standard error. Moreover, the regression equation relating 1963 to 1958 or 1954 per capita value added for the eighty-four large metropolitan areas is a good estimator of 1963 per capita value added in the few metropolitan areas that now fall in the size class, but were much smaller in the 1940s. For Bakersfield, Albuquerque, Orlando, San Bernardino, and Tucson, the mean 1963 per capita value added stood at $425; the corresponding mean estimated from the regression of 1963 on 1958 per capita value added is $414. For the latter four cities, the mean 1963 per capita value added was found to be $464; an estimate of $456 is obtained from the regression of 1963 on 1954 per capita value added over the eighty-four metropolitan areas. (We find that the regression coefficient is changed little when the interannual regression is calculated over areas as delimited at the date of the initial observation rather than over areas as delimited at the respective observation dates.) At least among larger communities, no systematic relation obtains between changes in the strength of the manufacturing sector in the local economy and changes in population size over a short interval of time.

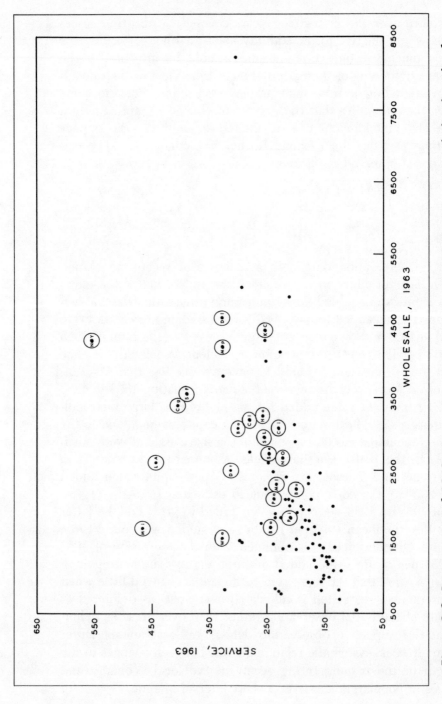

Figure 9-3. Scattergram of 1963 per capita service receipts on 1963 per capita wholesale sales for metropolitan areas (97 areas with 1960 populations of 250,000 or more).

Rapid population growth during the intercensal period had only a very slight positive effect on the level of per capita wholesale sales at the terminal date, net of the effect of the level of per capita wholesale sales at the initial date. The net regression coefficients, in standard form, associated with the population-growth measure in the four intercensal periods are: 1948 on post-1939 growth, .12; 1954 on post-1948 growth, .02; 1958 on post-1954 growth, .06; and 1963 on post-1958 growth, .04. The net regression coefficient associated with level of per capita sales at the initial date takes on a value of .98 or .99 in each period, by comparison. The relation of terminal level to recent growth and initial level, respectively, for service activity is similar to that just reported for wholesaling, that is, a very modest positive net effect associated with the population-growth measure, and a very strong positive net effect associated with initial level.

The scatter of points when per capita levels of manufacturing are plotted in relation to per capita levels of wholesaling suggests that the association between levels is essentially nil over large communities (for example, Figure 9-1). In contrast, the scatter of points when per capita levels of service receipts are plotted in relation to per capita levels of wholesaling suggests an average tendency for centers with high per capita wholesale levels to have also a high per capita level of service activity (Figure 9-3). The association between the per capita measures at a given point in time, in fact, is found to have been rather stable in recent decades. The coefficients of correlation measuring these associations are:

	1939	1947–1948	1954	1958	1963
Manufacturing with—					
wholesaling	−.23	−.15	−.02	−.05	−.09
services	−.22	−.18	−.05	−.07	−.06
Wholesaling with					
services	.65	.35	.59	.56	.54

A sophisticated modeling of changes over time in per capita levels of activity in the several sectors of the local economy might bring to light an effect of change in one sector on subsequent change in another sector. A superficial analysis suggests that such a modeling will not be a simple task, however. The per capita level of wholesaling does not appear to be influenced in any consistent

way by the per capita levels in manufacturing and service at some
prior date, net of the per capita level of wholesaling at that prior
date. The same finding can be reported with respect to per capita
levels of manufacturing and service activity. Shown below are the
partial regression coefficients in standard form obtained when sales,
value added, or receipts in 1963 was related to the three economic
measures in 1954 in a multiple-regression analysis; a parallel analysis is shown in which each economic measure in 1963 was related
to the three economic measures in 1939.

	1954			1939		
	Sales	Value	Receipts	Sales	Value	Receipts
Sales, 1963	.94	−.05	.03	1.01	.05	−.15
Value, 1963	.03	.91	−.03	.05	.95	.01
Receipts, 1963	−.06	−.03	.94	.05	.09	.80

Another way of summarizing the results is to report that a
prediction of 1963 per capita sales on the basis of 1954 per capita
sales alone would have been as successful (on the criterion of explained variance) as the corresponding prediction based on the
additive net effects of 1954 per capita sales, value added, and receipts. The same statement can be made when 1963 value added or
receipts is related to the 1954 economic measures or when 1963
sales or receipts is related to the 1939 economic measures. The only
instance in which improved prediction can be detected occurs when
1963 value added is related to the 1939 economic measures (a
fourteen-point increase in the proportion of explained variance).

Taken as a whole, the foregoing results suggest: first, that a
description of centers which serve as key points in the production
or distribution of goods and services within the nation is not likely
to be outdated for some years; and second, there is ample justification for analyzing the territorial organization of the manufacturing
sector independently of the territorial organization of the commercial sector. Moreover, although a coincidence of strength in several
sectors of the economy is more characteristic of the major metropolitan complexes than of lesser places, the relations between levels
of activity in the several sectors are sufficiently loose that differentiation among centers in their industrial structures are sure to be
observed.

STRUCTURAL DIFFERENTIATION
AMONG CENTERS

A set of statistics on the distribution by industry of the 1940 work force in metropolitan areas as delimited in 1965 which recently became available makes possible some comparisons among centers with respect to industrial structure. (The data were made available by the Office of Business Economics, U.S. Department of Commerce, and are the basis of their recent reports on "Growth Patterns in Employment by County 1940–50 and 1950–60.") In some centers, most notably those which were to experience substantial growth between 1940 and 1965, a sizable proportion of the work force was engaged in natural resource extraction or in construction in 1940. These categories along with "industry not reported" are excluded from the work force at the outset on the assumption that they are sensitive to the delimitation of areas and the local growth rate rather than to the "economic base" of the center.

With the industrialization of manufactures had come a distinctiveness in the industrial structures of the established centers which persisted in 1940 (Figure 9-4). Taking the nation's first city of New York as a baseline, the structures of the northeastern centers were without exception heavily weighted with manufactures and deficient in wholesaling, financial, and business activity and in the group of services here called personal and pleasure (entertainment, recreation, hotels, and "other" personal services). The centers to the west or south of the northeastern manufacturing belt were, on the other hand, without exception deficient in manufactures. Only New Orleans and Washington among the older centers and Houston and Miami among the new centers were deficient in wholesaling, financial, and business activity; and Minneapolis was the only center outside the northeast which was deficient in the personal-and-pleasure services.

Although the differences among the major centers with respect to these selected aspects of industrial structure may appear large, the comparisons between major centers and lesser places with respect to the mix of manufacturing and wholesaling suggested that the major centers had a more balanced industrial structure than the lesser places. Some additional support for this view is offered by an

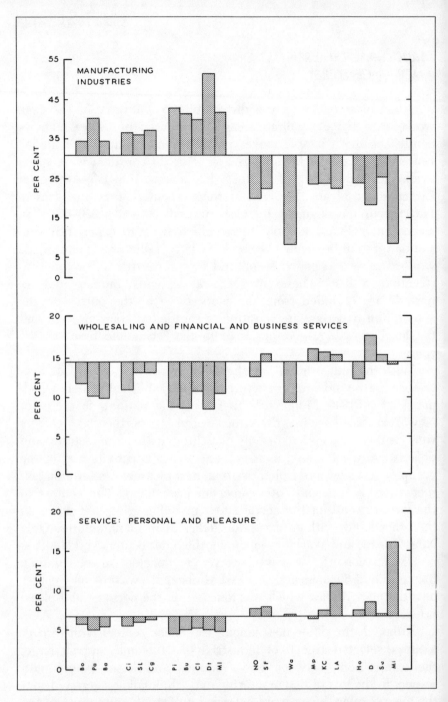

Figure 9-4. Percentage of employed in selected industries, for major centers in 1940 (Plotted as a deviation from the corresponding percentage in New York).

analysis of the dissimilarity in industrial structure between pairs of places. An index of dissimilarity with respect to the industrial distribution of the work force over thirty-one industry categories was calculated for 1, 596 pairs of places. (The places included the fifty-seven largest communities of 1960. The index equals half the sum of the absolute percentage-point differences between the distributions.) The mean index of dissimilarity for all pairs of places in 1940 was found to be 26.3 with a standard deviation of 8.4. For the eighteen largest centers of 1940, the mean index of dissimilarity in 1940 was found to be 21.2 with a standard deviation of 8.2. If the special case of Washington is excluded, the mean falls to 19.5 and the standard deviation to 6.8. The differences in industrial structure among the established centers averaged, then, somewhat less than the differences in industrial structure among a broader group of places.

Differentiation in industrial structure is patterned among the largest centers themselves. The centers of the northeast which had become established before the industrialization of manufactures resembled one another rather closely with respect to industrial structure. The centers outside the northeastern manufacturing belt also resembled one another rather closely with respect to industrial structure. The centers whose rise coincided with the industrialization of manufactures and which were from the outset centers of large-scale manufactures were rather dissimilar in industrial structure from one another as well as from the other groups of centers. Shown below are the mean indexes of dissimilarity for the groupings of centers:

		(1)	(2)	(3)
(1)	Bo, NY, Pa, Ba, Ci, SL, Cg	14	20	18
(2)	Pi, Bu, Cl, Dt, Ml	...	21	27
(3)	NO, SF, Mp, KC, LA	...	...	14

The pattern of differentiation among major centers is consistent with the notion that traditional metropolitan functions were combined with large-scale manufactures in the profiles of the older northeastern centers (Group 1). Their average dissimilarity from the centers of large-scale manufactures in the northeast (Group 2) was of about the same magnitude as their average dissimilarity from the trade-service centers to their south and west (Group 3) and was substantially less than the average dissimilarity of the northeastern manufacturing centers from the southern and western trade-service centers (Group 2 vs. Group 3).

To summarize the overall pattern of differences in industrial structure among major centers and especially to describe it in the context of differentiation among a broader group of centers is a near impossible task. We have calculated two measures of the difference in industrial structure, as it is revealed by a thirty-one-industry classification, between 1,596 pairs of sizable metropolitan areas for 1940, 1950, and 1960. The first measure is the index of dissimilarity; the second is the vector angle cosine. These alternative measures are closely, but imperfectly related, with the coefficient over the 1,596 observations taking on values of −.93 in 1940 and 1960 and −.94 for 1950. The degree of persistence in intercenter structural differences over ten-year spans is found to be high whether the difference is measured by the index of dissimilarity or the vector angle cosine. Using the former measure, the correlation coefficient is found to be .92 for 1940–50 and .94 for 1950–60 over the 1,596 observations or pairs of centers. The corresponding values are .88 and .93 when the vector angle cosine is used as the measure of difference in industrial structure.

The apparent temporal stability of differentiation and the insensitivity of results to the measure of structural difference suggest that a description based on one data set is not likely to lead to unwarranted generalization. A graphic representation of the matrix of intercenter differences as of 1960 is offered as a descriptive aid (Figure 9-5). The graphic representation is based on results of the Guttman-Lingoes "smallest space" analysis of the matrix; in the configuration, the "distance" of each center from each other center approximates the corresponding "distance" in the matrix, measured as rank order with respect to the vector angle cosine, as closely as possible in a two-space.

Ten major centers can be located in a small segment of the two-space which is occupied by no lesser centers. This segment is, moreover, positioned centrally in the overall configuration. Given our perspective on the development of a system of differentiated centers, we could reason along the following lines in interpreting this feature of the configuration.

Boston, New York, Philadelphia, Baltimore, New Orleans, Cin-

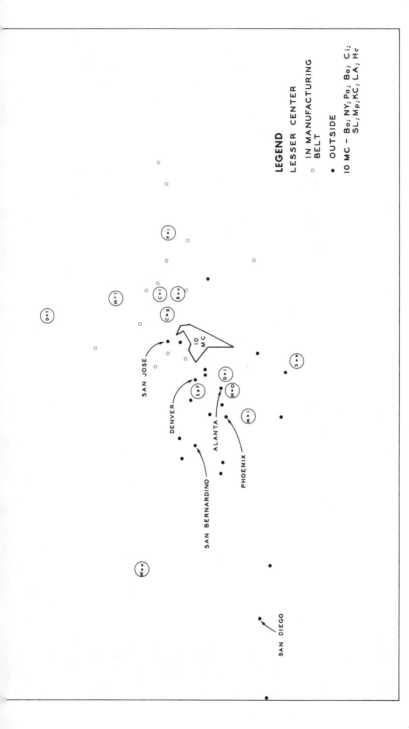

Figure 9-5. Dissimilarity in industrial structure among major centers and 35 lesser centers in 1960.

cinnati, and St. Louis were well-established centers of commerce
at a time when the national economy could appropriately be conceived as a mosaic of resource regions linked through their respective commercial capitals. Chicago was the newest such center
when the economy began to be transformed by the industrialization
of manufactures, and San Francisco was emerging as the commercial capital of the Pacific resource region whose exploitation
was gaining momentum. Large-scale manufactures were, in a sense,
appended to the established economic core in all these commercial
capitals save New Orleans and San Francisco which were isolated
from the mainstream of settlement.

In the representation of diversity in industrial structure as of
1960, Boston, New York, Philadelphia, Baltimore, Cincinnati, and
St. Louis occupy positions in the centrally located segment of the
two-space. Chicago occupies a position somewhat closer to the
five trans-Appalachian centers whose rise to prominence coincided
with the industrialization of manufactures, and San Francisco and
New Orleans occupy positions more distant from the manufacturing centers.

The remaining positions in the centrally located segment of
the two-space are occupied by Minneapolis, Kansas City, Los
Angeles, and Houston, centers whose rise to prominence postdates
the settlement of the national territory and the appearance of industrialized manufactures. Resources and markets reinforced the
advantage of the northeastern part of the nation as the site of a
manufacturing belt within which specialized manufacturing centers
had emerged. Centers to the south or west could attain great size
neither as a manufacturing center nor as the capital of a resource
region, but only by mixing manufactures and commerce as had
the older centers that ringed the manufacturing belt.

The three newest of the major centers do not occupy positions
in the centrally-located segment of the two-space, nor does Washington. The latter, we have argued, is a special case whose industrial structure could scarcely be replicated by another center except
in the event of a marked decentralization of activities of the
federal government. The positions of Dallas, Seattle, and Miami
pose a more interesting problem, however; for these centers are the
first to gain prominence since World War II. They now occupy
positions closer to that of New Orleans, once a collecting point in

a colonial resource region, than of any other major center. Before speculating on the significance of their positions, it may be well to examine the configuration of lesser centers.

Identification of the lesser centers appearing in the representation according to whether each lies within or outside the northeastern manufacturing belt reveals relatively little overlap in the positions of the two groups of centers. Lesser manufacturing-belt centers typically lie closer to one of the five major centers of manufacture than to any other major center; their counterparts outside the manufacturing belt typically lie closer to, say, San Francisco or New Orleans.

The instances of overlap are themselves instructive. Birmingham accounts for the sole instance in which a lesser center outside the manufacturing belt lies closer to a major center of manufacture than to any other center; a pre-1900 localization of the iron and steel industry appears in Alabama where ore and coal deposits were found in proximity. All other instances of overlap are concentrated within a relatively small, centrally located segment of the two-space occupied by Albany, Columbus, Indianapolis, Louisville, and Richmond, and San Jose, a rapidly growing center of the far west. The importance of Albany as a transshipment center prior to the industrialization of manufactures has been documented in Chapter 3; and like older major centers of the northeast, Albany reveals a contemporary industrial structure resembling a blend of the structures characteristic of a manufacturing center and a commercial capital. A case could be made for identifying Columbus and Indianapolis, which lie within the manufacturing belt, and Louisville and Richmond, which lie just below the belt's southern edge as it is usually demarcated, as focal points in encircling resource regions before heavy manufactures became a national growth industry.

Some centers which we suggest may soon be contenders for status as major centers given their current populations and recent growth rates are Atlanta, Denver, and San Diego. Smaller, but shifting upward in rank rapidly, are San Jose, San Bernardino, and Phoenix. These contenders for national prominence differ sharply from one another in industrial structure, but, with the exception of San Jose, each occupies a position in the configuration closer to San Francisco or New Orleans than to the other major centers in which manufactures are relatively prominent in the industrial mix.

In this respect, they resemble the newest of the major centers—Dallas, Seattle, and Miami.

Again we must pose the problem: are the new centers to remain distinctive in industrial structure, or are we capturing their structures in a transitional stage. The evidence for short-run stability in function is unambiguous. Working from our 1900 bench mark, we can now assess long-run stability with respect to two major functions of the major centers.

Chapter **10**

MANUFACTURING PROFILES IN MAJOR CENTERS

In the six decades following 1900, the national industrial structure underwent a marked shift away from resource extraction. Primary resource extractors, which in 1900 had employed perhaps two-fifths of the national work force, provided employment for only 14 percent of the 1950 labor force and 8 percent of the 1960 labor force. The proportions of the national work force engaged in branches of manufacture classified as first-stage or second-stage resource users stood at about the same level in 1950 or 1960 as in 1900. First-stage users provided employment for 4 percent of the work force, and second-stage users provided employment for 8 percent. Gains in employment in the manufacturing sector had been concentrated heavily in industries which were less directly dependent on natural resource extraction, and their share of the national work force rose from 5 percent in 1900 to 13 percent in 1950 and 15 percent in 1960. Among the major employers in this expanding industry category were manufacturers of motor vehicles and aircraft, unambiguously twentieth-century lines of manufacture. (The basic data for 1950 and 1960 are from the Censuses of Population conducted in the respective years.)

The level of specialization in manufactures, as indexed by the location quotient, was less pronounced at mid-century than in

1900 for the group of major centers for which a 1900 bench mark was established. We estimate the mean location quotient to have been 1.6 in 1900, in contrast to the 1.23 recorded in 1950 or the 1.14 recorded in 1960. The decrease in the share of the national work force engaged in resource extraction has been so substantial since 1900, however, that if the national nonextractive work force is taken as the norm of comparison, manufacturing workers are as heavily overrepresented in the major centers now as they were in 1900.

The change in mean level of specialization for the major centers varies markedly among resource-market categories of manufacture. The lines in which the established centers remain most clearly differentiated from the rest of the nation are those that we referred to earlier as nontraditional, that is, industries relying on other processors for inputs and marketing their products to other industrial establishments. The distinctiveness of the major centers with respect to other kinds of manufacture, most notably second-stage resource users producing for the final market, has been lost, however.

From the perspective of localization, the stability of established centers as sites of manufacturing activity is noteworthy. Concentrated in New Orleans and the fifteen centers with populations exceeding that of New Orleans in 1900 were 33 percent of the national manufacturing work force at the turn of the century, 35 percent of the national manufacturing work force as of 1960. (A center's territory is defined as the metropolitan area in 1960, as the center and large cities in what now is the metropolitan area in 1900. Only insofar as sizable manufacturing establishments were clustered in small towns and open country proximate to the center in 1900 is the estimate of change influenced by territorial redefinition. Moreover, a case could be made that engulfment of outlying industrial concentrations is a normal part of the growth of a center.)

The aggregate share of the established centers had been at a maximum for second-stage resource users producing for the final market in 1900, but in 1960 their share reached a maximum for the "indirect, nonfinal" category. Reflected are a very sharp drop in the concentration of "second-stage, final" manufactures and a very modest increase in the concentration of "indirect, nonfinal" manufactures. Less substantial absolute losses, in terms of percentage share, were recorded for first-stage resource users and in the indirect-final category. A more substantial gain was recorded for second-stage users producing for the nonfinal market. The holding, or attracting, power of the major centers was stronger in the nontraditional lines of urban manufacture than in lines which might be an outgrowth of a center's metropolitan status (Table 10-1).

In 1960, as in 1900, the largest single concentration of manufacturing activity was to be found in New York and its environs; the second largest concentration again was to be found in Chicago and its environs; still occupying third rank was Philadelphia. With respect to share of the manufacturing total, Chicago had moved a little closer to New York, a little further from Philadelphia. The rank order was not in question, however, for New York's share in 1960 stood at 9.9 percent, in contrast to 5.5 percent for Chicago and 3.4 percent for Philadelphia. (By 1960 the newer center of Los Angeles with 4.6 percent of the nation's manufacturing workers had broken into the ranks of the established "top three.") Boston, the fourth-ranking center of manufacture in 1900, had been displaced by both Detroit and Pittsburgh; Pittsburgh, in turn, had been displaced by Detroit whose share had grown to 3.1 percent.

Ranking first with respect to share of the national total in each category of manufacture in 1960 was New York, and its pre-eminence represented no change from the 1900 bench mark. National prominence in one resource-market category was less likely to ensure prominence in other categories at mid-century than in 1900, however. Chicago would have been counted six times, Philadelphia five times, and Boston once in an enumeration of centers occupying the second and third ranks in 1900. Chicago would appear in the 1960 roster six times (although the newer center of Los Angeles now had greater shares in the "second-stage, final" and "indirect, final" categories). Philadelphia would appear but three times, for first-stage resource users and second-stage users producing for the final market; and included in the roster would be: Pittsburgh for second-stage users producing for the nonfinal market; Detroit for the "indirect, nonfinal" category; and St. Louis for the "indirect, final" category. In this particular sense, differentiation of role among centers of manufacture has become sharper.

Overall, the established centers of 1900 are as important in the

Table 10-1. Percentage Share of the 1960 National Manufacturing Work Force in Major Centers of 1900, and Change in Share over 1900, for Industry Categories Defined by Stage of Resource Use and Type of Market

Center	Manufac- turing, All	FIRST STAGE Non- final	Final	SECOND STAGE Non- final	Final	INDIRECT Non- final	Final
		Percentage Share 1960					
All	35.04	16.75	29.26	31.58	36.25	40.72	29.28
Boston	1.69	0.36	1.55	0.70	1.56	2.23	2.11
New York	9.93	3.74	7.20	6.01	19.32	9.66	13.85
Philadelphia	3.37	2.76	3.01	3.09	4.33	3.72	1.34
Baltimore	1.11	0.45	1.11	1.67	1.07	0.97	1.30
Cincinnati	0.75	0.21	0.74	0.73	0.58	0.78	1.31
St. Louis	1.44	1.06	1.58	1.30	1.00	1.42	2.75
Chicago	5.46	3.32	4.92	5.87	3.28	6.77	2.58
Pittsburgh	1.76	0.43	1.09	4.51	0.52	1.63	0.26
Buffalo	1.03	0.62	0.47	1.96	0.53	1.03	0.69
Cleveland	1.57	0.38	0.67	1.72	0.85	2.11	1.07
Detroit	3.09	0.43	1.28	1.63	0.74	5.39	0.66
Milwaukee	1.09	0.15	1.18	0.62	0.40	1.69	0.43
New Orleans	0.27	0.35	0.75	0.29	0.31	0.21	0.06
San Francisco	1.28	1.23	2.09	1.00	0.97	1.49	0.43
Minneapolis	0.86	1.14	1.26	0.37	0.60	1.09	0.32
Washington	0.34	0.12	0.36	0.11	0.19	0.53	0.12
		Percentage-Point Change Over 1900					
All	2.0	—4.4	—10.5	10.5	—19.0	4.0	—12.2
Boston	—0.2	—1.1	—0.2	0.1	—0.5	—0.6	—3.9
New York	—0.9	—2.2	—3.9	0.7	—4.8	—1.5	—2.4
Philadelphia	—1.4	0.2	—0.5	—1.4	—3.8	—0.7	—2.8
Baltimore	—0.2	—0.2	—1.6	1.2	—2.2	—0.1	0.0
Cincinnati	—0.4	—0.2	—0.9	0.1	—1.2	—0.5	—1.5
St. Louis	0.0	0.1	—0.9	0.6	—0.8	—0.3	0.6
Chicago	0.4	0.5	—4.0	3.2	—3.4	—0.2	—1.7
Pittsburgh	0.1	—1.2	—0.2	2.1	0.0	—0.1	0.1
Buffalo	0.4	—0.1	—0.3	1.5	—0.2	0.4	0.1
Cleveland	0.6	—0.3	0.0	0.6	—0.2	0.7	0.7
Detroit	2.4	0.0	0.2	1.1	—0.2	4.5	0.0
Milwaukee	0.2	—0.4	0.0	0.0	—0.7	0.7	—0.4

Table 10-1. (continued)

Center	Manufac- turing, All	FIRST STAGE Non- final	Final	SECOND STAGE Non- final	Final	INDIRECT Non- final	Final
	Percentage-Point Change Over 1900						
New Orleans	0.0	0.2	0.1	0.1	—0.4	0.0	—0.2
San Francisco	0.5	0.6	0.8	0.6	—0.3	0.8	—0.3
Minneapolis	0.3	0.0	0.7	0.0	—0.3	0.6	—0.5
Washington	0.2	—0.3	0.2	0.0	0.0	0.3	0.0

territorial organization of manufacturing activity now as six decades ago although major concentrations have developed elsewhere in the nation, notably in Los Angeles. The new manufacturing concentrations represent a net redistribution of manufacturing activity away from territory outside the established centers, not from the established centers themselves. This is not to say that the manufacturing specialties of the established centers remain unchanged. Indeed, these centers could maintain their shares of the nation's manufacturing activity only by developing new specialties or by capturing a still larger share of activity in the lines of manufacture that first had developed as local specialties.

PERSISTENCE IN ESTABLISHED CENTERS

Distinct from the manufacturing specialization of the established centers relative to the rest of the nation is their specialization relative to one another. With the evidence at hand, one could not argue that differences among the nation's major centers with respect to industrial structure had become increasingly blurred over time. Although, on at least one criterion of variability, the centers became slightly less heterogeneous with respect to level of specialization in the manufacturing sector as a whole, the recorded decrease for all manufactures is small; and for some resource-market categories increases, not decreases, in variability are recorded. (The criterion is the ratio of standard deviation to mean measured with respect to the location quotient and reported in Table 10-2). The limited range within which the ratios for any given category

Table 10-2. Interannual Regressions of Location Quotients, 1900 to 1960, Over Major Centers of 1900

Year(s)	Manufac- turing, All	FIRST STAGE Non- final	Final	SECOND STAGE Non- final	Final	INDIRECT Non- final	Final
				Mean			
1960	1.14	0.57	1.05	1.18	0.84	1.36	0.86
1950	1.23	0.62	1.23	1.14	0.97	1.52	1.01
1900	1.64	1.21	2.13	1.15	2.36	1.85	1.90
			Standard Deviation/Mean				
1960	.29	.61	.36	.72	.56	.42	.80
1950	.32	.53	.43	.76	.61	.51	.62
1900	.35	.40	.44	.70	.47	.38	.71
			Correlation Coefficient				
1960 on 1950	.99	.88	.95	.99	.98	.98	.91
1960 on 1900	.83	.25	.41	.72	.70	.74	.75
1950 on 1900	.79	.24	.52	.75	.70	.67	.67
			Regression Coefficient				
1960 on 1950	.83	.92	.67	.98	.78	.72	1.01
1960 on 1900	.49	.17	.16	.75	.30	.60	.38
1950 on 1900	.56	.16	.29	.81	.38	.73	.31

Note: Centers are identified in Table 10-1.

fluctuate and the temporal stability of cross-sectional differences in variability by resource-market category are the salient features of the comparison.

The relative positions of the established centers with respect to level of specialization in the manufacturing sector as a whole have changed relatively little since the turn of the century. When the 1960 or 1950 location quotient for all manufactures is regressed on the corresponding 1900 location quotient, a correlation coefficient of .8 over the sixteen major centers of 1900 is obtained. The corresponding regression coefficient is on the order of .5, indicating an average tendency for centers with initially low levels of manufacturing specialization to experience the more substantial gains in the ensuing period.

Specialization in first-stage industries appears to be transitory, for the mid-century level of specialization varies over centers independently of their respective levels at the turn of the century. Specialties in the kinds of manufacture less directly dependent on resource extraction for their inputs, in contrast, tend to persist. The association of initial and terminal levels of specialization is about as close for second-stage and indirect users producing for the final market as for the nontraditional branches in which both suppliers and customers are typically other industrial establishments. It is in the nontraditional branches of manufacture, however, that the advantage of an initially high level of specialization has been maintained most effectively (a point revealed by the relatively large regression coefficients).

A model assuming random disturbances which alter and eventually destroy the initial pattern of intercenter differences does not seem to hold promise in accounting for these observations on temporal stability. For all manufactures and for three of the six resource-market categories, the coefficient of correlation between 1900 and 1960 location quotients is larger in magnitude than the coefficient between 1900 and 1950 quotients. For the other resource-market categories, the 1900–1960 coefficient is no larger than the 1900–1950 coefficient, but the product of the 1900–1950 and 1950–1960 coefficients, respectively, is unreliable as an estimator of the value of the 1900–1960 coefficient. Moreover, assuming a simple chain model of change and constant coefficients at the 1950–1960 value over each ten-year span since the turn of the century, the 1900–1960 coefficient is over-estimated for some categories and underestimated for other categories. The industrial distribution of the work force at an arbitrarily selected point in time may be a fallible indicator of the average industrial distribution over, say, a two-year period centered on the measurement date, of course; and our readings on the industrial structure are so few that we cannot distinguish "fluctuations" from "alterations" in pattern.

Cursory inspection of the location quotients for all manufactures in 1960 reveals that manufacturing workers are overrepresented in each of the twelve established centers whose configuration had defined the manufacturing belt of 1900. The location quotient takes on higher values in Pittsburgh, Buffalo, Cleveland, Detroit, and Milwaukee, however, than in the older metropolitan centers which ring them. As late as 1960, only Los Angeles among the major centers to the south or west of the manufacturing belt had a disproportionate share of its work force in the manufacturing sector (Table 10-3).

Short-run stability in the level of manufacturing specialization

Table 10-3. 1960 Location Quotients in Six Manufacturing Industry Categories, and Change over the Corresponding Quotient in 1950, for Major Centers of 1920

Center	Symbol	Manufacturing, All	FIRST STAGE Non-final	Final	SECOND STAGE Non-final	Final	INDIRECT Non-final	Final
				Quotient, 1960				
Boston	Bo	1.08	0.23	0.99	0.45	1.00	1.43	1.35
New York	NY	1.09	0.41	0.79	0.66	2.12	1.06	1.52
Philadelphia	Pa	1.33	1.09	1.19	1.22	1.71	1.47	0.53
Baltimore	Ba	1.13	0.46	1.13	1.70	1.09	0.99	1.33
Cincinnati	Ci	1.24	0.35	1.22	1.21	0.96	1.30	2.17
St. Louis	SL	1.24	0.91	1.36	1.12	0.86	1.22	2.37
Chicago	Cg	1.33	0.81	1.20	1.43	0.80	1.65	0.63
Pittsburgh	Pi	1.36	0.33	0.84	3.48	0.40	1.26	0.20
Buffalo	Bu	1.40	0.84	0.64	2.65	0.72	1.40	0.93
Cleveland	Cl	1.46	0.35	0.62	1.60	0.79	1.96	0.99
Detroit	Dt	1.50	0.21	0.62	0.79	0.36	2.62	0.32
Milwaukee	MI	1.50	0.20	1.62	0.85	0.55	2.32	0.59
New Orleans	NO	0.59	0.77	1.64	0.63	0.67	0.46	0.14
San Francisco	SF	0.78	0.75	1.27	0.61	0.59	0.91	0.26
Minneapolis	Mp	0.96	1.28	1.41	0.41	0.67	1.22	0.36
Kansas City	KC	0.93	1.12	1.13	0.70	0.87	1.04	0.50
Los Angeles	LA	1.14	0.66	0.72	0.48	0.98	1.23	3.12
Washington	Wa	0.28	0.10	0.30	0.09	0.16	0.44	0.10
				Change Over 1950				
Boston	Bo	0.0	—0.1	0.0	—0.1	—0.1	0.0	—0.3
New York	NY	—0.1	—0.2	0.0	0.0	—0.4	0.0	—0.4
Philadelphia	Pa	—0.1	—0.1	—0.1	0.0	—0.4	0.0	—0.3
Baltimore	Ba	0.0	—0.1	—0.2	0.2	0.0	—0.1	0.0
Cincinnati	Ci	0.0	—0.1	—0.1	0.2	—0.1	—0.3	0.6

Table 10-3. (continued)

Center	Symbol	Manufacturing All	FIRST STAGE Non-final	Final	SECOND STAGE Non-final	Final	INDIRECT Non-final	Final
			Change Over 1950					
St. Louis	SL	—0.1	0.0	—0.7	0.1	—0.2	—0.1	0.2
Chicago	Cg	—0.1	0.1	—0.6	0.1	—0.2	—0.2	—0.5
Pittsburgh	Pi	—0.1	—0.1	0.0	—0.2	0.1	—0.1	—0.1
Buffalo	Bu	—0.2	—0.4	—0.1	0.1	0.0	—0.3	—0.2
Cleveland	Cl	—0.2	0.0	0.0	0.0	—0.1	—0.3	—0.3
Detroit	Dt	—0.3	—0.1	0.0	0.1	0.1	—1.0	—0.1
Milwaukee	Ml	—0.2	0.0	—0.5	0.0	—0.3	—0.3	—0.5
New Orleans	NO	0.0	0.2	—0.2	0.0	—0.2	0.0	0.0
San Francisco	SF	0.0	—0.2	0.0	0.1	0.0	0.0	0.0
Minneapolis	Mp	0.0	0.3	—0.2	0.0	—0.2	0.0	—0.5
Kansas City	KC	0.0	0.1	—0.7	0.0	—0.1	0.1	0.0
Los Angeles	LA	0.2	—0.2	—0.2	0.1	—0.2	0.3	0.4
Washington	Wa	0.0	0.0	0.0	0.0	0.0	—0.1	0.0

in major centers can be inferred from the typically negligible changes in location quotients recorded between 1950 and 1960. Three of the five most sizable changes recorded represent decreases in the location quotient for "first-stage, final" industries in St. Louis, Chicago, and Kansas City. That this kind of specialty was relatively vulnerable to change had been apparent by the turn of the century. A decrease for "indirect, nonfinal" industries in Detroit and an increase for "indirect, final" industries in Cincinnati are the only other changes of equal magnitude.

If the configuration of specialties by resource-market category can convey the "character" of local manufactures, there has been detectable continuity over time in such character. On the other hand, there is little support for the notion that specialization in a given resource-market category predisposes a center to subsequent specialization in another category. Some more or less typical evolution of the local industrial structure may accompany the transformation of a small community into a leading center, perhaps along the lines suggested by Thompson (1965: 434–445); but, if so, the patterned industrial change has come to an end by the time a place enters the ranks of the leading centers. A distinctive structure then has developed which tends to persist thereafter.

In the cross-section, the level of specialization in one kind of manufacture tends to vary over centers independently of the level of specialization in any other kind of manufacture. Four marginally significant positive associations between resource-market categories are to be observed in 1900; one such association is to be observed in 1950; none is found in 1960. (Coefficients of correlation ranging from .50 to .65 obtain between the following pairs of location quotients over the 16 major centers of 1900: first-stage and second-stage users producing for the nonfinal market in 1900; first-stage and second-stage users producing for the final market in 1900; first-stage users producing for the final market and indirect users producing for the nonfinal market in 1900; second-stage and indirect users producing for the nonfinal market in 1900; second-stage and indirect users producing for the final market in 1950.)

Only three lagged associations between level of specialization in one kind of manufacture and level of specialization in another kind are detected. Centers in which the level of specialization in "first-stage, nonfinal" manufactures was relatively high in 1900 have a relatively high level of specialization in "second-stage, nonfinal" manufactures in both 1950 and 1960. Interesting as such an association might be from the standpoint of industrial evolution, it appears to be accounted for adequately by the initial linkage between specializations in "first-stage, nonfinal" and "second-stage, nonfinal" industries and the persistence through time of "second-stage, nonfinal" specialties. (The relevant coefficients of correlation are: 1NF, 1900 with 2NF, 1950 = .55; 1NF, 1900 with 2NF, 1960 = .52; 1NF, 1900 with 2NF, 1900 = .56; and the interannuals reported in Table 10-2). The second instance of lagged association seems substantively uninteresting, as well as marginally significant in a statistical sense: indirect users producing for the final market in 1950 with second-stage users producing for the final market in 1960.

Most major centers would be characterized in much the same way now as in 1900 with respect to the importance of center-region linkages for local manufactures. The specialties of New Orleans and San Francisco, for example, have been exclusively first-stage or second-stage resource users producing for the final market, traditional specialties reflecting the role of the center as metropolis for a resource region. In contrast, the nontraditional branches of manufacture oriented to other industrial establishments rather than to resource extractors and final consumers have been the specialties of Pittsburgh and Cleveland. Discontinuities in the character of local manufactures are perhaps most evident in Baltimore, Buffalo, and Detroit where regionally-oriented specialties have waned and in Minneapolis where regionally-oriented specialties have become more salient (Table 10-4).

A variety of comparisons based on change in resource-market specialties are possible. Aside from continuity in character, we single out for comment only the "disappearance" of one kind of specialty, however. In 1900 Pittsburgh and Washington were distinguished by an absence of specialization with respect to second-stage users producing for the final market; by 1960 New York and Philadelphia had become distinguished by the presence of this specialty. The specialty had disappeared from the industrial profiles of twelve of the sixteen established centers in the intervening six decades. Not to be ruled out in accounting for the recurrent disappearance is the possibility that only in major centers had producers become differentiated from consumers for such goods as clothing or furniture by 1900. In this event, the loss of the specialty would reflect convergence between large-city and less urban populations in the range of goods secured in the market, rather than change in a center-region relation.

Thus far we have observed a rather remarkable stability over a sixty-year span in the ranking of centers with respect to level of specialization in the manufacturing sector as a whole and an appreciable continuity in the character of local manufactures in terms of resource use and market type. When the detailed industrial profile in the manufacturing sector is examined for each center, both losses and gains of local specialties are almost certain to be recorded; but these changes have balanced in such a way that the center typically has retained its distinctiveness vis-à-vis other centers.

A first impression upon inspecting the 1960 industrial profiles in the manufacturing sector may be that they are less full or diversified than the 1900 profiles. The mean number of specialties

Table 10-4. Rank of Resource-Market Manufacturing Specialties in Each Major Center of 1900: 1900, 1950, and 1960 [— indicates location quotient under 1.1; ( ) indicates location quotient 1.1 to 1.7]

Center	1NF			1F			2F			1F			1NF			2NF	
	1900	1950	1960	1900	1950	1960	1900	1950	1960	1900	1950	1960	1900	1950	1960	1950	1960
Bo	—	—	—	(4)	—	—	(3)	(3)	—	1	(1)	(2)	2	(2)	(1)	—	—
NY	(6)	(5)	—	4	—	—	1	1	1	2	2	(2)	3	—	—	—	—
Pa	—	—	—	(5)	(3)	(4)	1	1	1	4	(1)	—	3	(2)	(2)	(4)	(3)
Ci	(5)	—	—	3	(3)	(3)	2	(4)	—	1	(1)	1	4	(2)	(2)	—	(4)
SL	(5)	—	—	1	2	(2)	3	(4)	—	2	1	1	4	(3)	(3)	—	(4)
Cg	—	—	—	1	2	(3)	3	—	—	4	—	—	2	—	(1)	(3)	(2)
Pi	3	—	—	4	—	—	—	—	—	—	—	—	2	(2)	(2)	—	1
Cl	(4)	—	—	(5)	—	—	2	—	—	—	(3)	—	—	—	—	(2)	(2)
Ml	(6)	—	—	1	2	(2)	2	—	—	4	(3)	—	3	—	—	—	—
NO	—	—	—	(2)	1	(1)	1	—	—	—	—	—	—	—	—	—	—
SF	—	—	—	1	(1)	(1)	2	—	—	(3)	—	—	—	—	—	—	—
Wa	—	—	—	—	—	—	—	—	—	—	—	—	—	—	—	—	—
Ba	—	—	—	2	(2)	(3)	1	(4)	—	3	(3)	(2)	(4)	—	—	—	—
Bu	(3)	(3)	—	1	—	—	(2)	—	—	(5)	(4)	—	(4)	(2)	(2)	(1)	1
Dt	—	—	—	(5)	—	—	2	—	—	(4)	—	—	3	—	—	—	1
Mp	1	—	(2)	(5)	(1)	(1)	(2)	—	—	(3)	—	—	(4)	(2)	(3)	—	—

Note: Caption and stub symbols are identified in Table 10-3.

164

(detailed manufacturing industries for which the location quotient takes on a value of at least 2.0) had fallen from 16.6 to 7.8 in the five manufacturing-belt centers which had attained prominence just before 1900 and evidenced the highest level of manufacturing specialization at mid-century. In the seven older metropolitan centers which ringed them, the number of specialties averaged 20.1 in 1900 as compared with 9.2 in 1960. (The 1960 profiles appear in Tables 10-5 and 10-6.)

Net changes in specialties within resource-market category sharpened the difference between the two groups of centers with respect to the importance of "metropolitan" manufactures. Shown below are the mean numbers of specialties by resource-market category at the two dates.

	Five centers		*Seven centers*	
	1900	*1960*	*1900*	*1960*
First stage	3.6	0.4	3.9	1.8
Second, final	2.4	0.0	3.4	0.5
Indirect, final	0.8	0.0	2.0	1.4
Nontraditional	9.8	7.4	10.8	5.5

The specialties of Pittsburgh, Buffalo, Cleveland, Detroit, and Milwaukee are now almost exclusively in lines of manufacture for which suppliers and customers are other industrial establishments, but the mix of traditional and nontraditional specialties in the seven older metropolitan centers has remained essentially unchanged.

Comparison of the 1900 and 1960 profiles for individual centers of the northeast points up the loss of all first-stage industries which had been both specialties and major employers in 1900 with the exception of the beverage industries in Cincinnati, St. Louis, and Milwaukee. A similar comparison for "second-stage, final" specialties points up the loss of all such specialties except the apparel industry in New York and the knitting-mills and floor-coverings industries in Philadelphia. An "indirect, final" specialty remains only in Boston and New York, Cincinnati and St. Louis. Instances of stability in detailed manufacturing specialties become more frequent within the nontraditional resource-market categories (Table 10-7).

Between 1900 and 1960, some new specialties developed in the established centers; or more accurately, lines of manufacture appear

Table 10-5. Industrial Profiles in the Manufacturing Sector for Boston, New York, Philadelphia, Baltimore, Cincinnati, St. Louis, and Chicago, 1960

Stage of Resource-Use, Type of Market, and Detailed Industry	LOCATION QUOTIENT FOR						
	Bo	NY	Pa	Ba	Ci	SL	Cg
First Stage, Nonfinal							
Misc. nonmetallic mineral and stone products	...ᵃ	...	...	...	...	...	2.1
Petroleum refining	...	...	3.3ᵇ	...	...	2.1	...
Grain-mill products	...	...	...	...	...	2.2	...
Dyeing and finishing textiles, except wool and knit	...	2.1	...	...	...	...	...
First Stage, Final							
Confectionary and related products	5.4	...	2.5	...	...	...	4.3
Beverage industries	...	...	...	2.1	3.3	3.2	...
Meat products	...	...	...	...	...	2.0	...
Second Stage, Nonfinal							
Paints, varnishes, and related products	...	...	...	2.0	...	...	2.9
Misc. textile mill products	...	...	2.3	...	...	...	...
Misc. petroleum and coal products	...	...	2.8	...	6.4	2.6	3.7
Blast furnaces, steel works, and rolling and finishing mills	...	...	...	5.6	...	...	3.4
Misc. chemicals and allied products	...	...	...	...	3.3	2.5	...
Misc. plastic products	...	...	...	...	2.5	...	2.4
Second Stage, Final							
Misc. fabricated textiles	...	2.9	...	...	...	...	...
Apparel and accessories	...	2.9	1.9	...	...	...	...
Floor coverings, except hard surface	...	...	4.6	...	...	...	...
Knitting mills	...	...	2.4	...	...	...	...
Indirect, Nonfinal							
Misc. machinery	...	...	...	...	1.9	...	1.7
Printing, publishing, and allied industries, except newspapers	...	2.3	...	...	2.4	...	2.6
Rubber products	2.8	...	...	...	...	...	...
Electrical machinery, equipment, and supplies	2.7	...	1.8	...	...	...	2.3
Drugs and medicines	...	3.2	2.9	...	...	...	...
Paperboard containers and boxes	...	...	2.3	2.6	...	...	...
Professional equipment and supplies	...	2.6	2.0	...	...	...	...
Leather: tanned, curried, and finished	11.3	...	2.1	...	...	...	...

Table 10-5. (continued)

Stage of Resource-Use, Type of Market, and Detailed Industry	Bo	NY	Pa	Ba	Ci	SL	Cg
Indirect, Nonfinal—cont.							
Misc. fabricated metal products	...	...	*1.7*	...	...	...	2.2
Misc. paper and pulp products	2.0	...	...	...	...	2.3	...
Ship and boat building and repairing	3.4	...	3.6	3.1	...	...	...
Railroad and misc. transportation equipment	...	...	...	...	...	3.2	3.7
Farm machinery	...	...	...	...	...	...	2.4
Glass and glass products	...	...	...	...	...	3.3	...
Photographic equipment and supplies	3.4	...	...	...	...	...	3.2
Motor vehicles and motor vehicle equipment	...	...	...	...	*1.9*	...	...
Indirect, Final							
Footwear, except rubber	3.6	...	...	...	...	3.6	...
Misc. manufacturing industries	...	2.5	...	...	2.4	...	...
Watches, clocks, and clockwork-operated devices	...	2.2	...	...	...	...	2.8
Leather products, except footwear	...	4.2	...	...	...	...	...
Aircraft and parts	...	...	...	2.2	2.8	2.7	...

a. ... indicates location quotient less than 2.0.
b. Italic figures indicate at least 1.0 percent of local work force in industry.

in the 1960 industrial profiles which were not present in the 1900 profiles. Some of these specialties—ship-and-boat building and re-pairing, for example—are probably not new to the community, but appear so only because of changes in industrial classification or territorial delimitation of centers. In a number of instances, how-ever, there is reason to think that the appearance of a new specialty in an established manufacturing center is not artifactual. Electrical machinery, for example, is by no means a new manufacturing specialty although it is new to Milwaukee where employment is concentrated in "industrial controls." An important component of the primary-nonferrous specialty in Pittsburgh is "aluminum rolling and drawing"; in Cleveland, an important component is "aluminum castings." There are grounds for thinking that the specialty as pursued in the center is new nationally as well as locally. The "new-ness" of the local specialty is even less in question when the line

Table 10-6. Industrial Profiles in the Manufacturing Sector for Pittsburgh, Buffalo, Cleveland, Detroit, and Milwaukee, 1960

Stage of Resource-Use, Type of Market and Detailed Industry	LOCATION QUOTIENT FOR				
	Pi	Bu	Cl	Dt	Ml
First Stage, Nonfinal					
Grain-mill products	...[a]	3.9	...	...	...
First Stage, Final					
Beverage industries	...	...	...	...	*5.1*[b]
Second Stage, Nonfinal					
Paints, varnishes, and related products	...	...	6.4	2.2	2.2
Misc. textile mill products	...	...	2.5	...	...
Primary iron and steel industries, except blast furnaces, etc.	*6.9*	...	*2.8*	...	*5.5*
Misc. petroleum and coal products	3.3	...	3.8	...	...
Blast furnaces, steel works, and rolling and finishing mills	*12.1*	*6.1*	*3.1*	...	...
Misc. chemicals and allied products	...	*3.5*	...	...	...
Misc. plastic products	...	3.2	...	...	...
Synthetic fibers	...	...	2.2	...	...
Indirect, Nonfinal					
Misc. machinery	...	...	2.7	2.5	4.0
Electrical machinery, equipment, and supplies	*1.9*	...	*1.6*	...	*3.1*
Professional equipment and supplies	...	...	...	...	*3.7*
Leather: tanned, curried, and finished	...	...	...	...	11.3
Misc. fabricated metal products	...	...	2.6	2.0	...
Fabricated structural metal products	3.3	...	...	...	...
Cutlery, hand tools, and other hardware	...	...	...	2.1	...
Railroad and misc. transportation equipment	3.4	...	...	...	...
Primary nonferrous industries	*2.9*	*2.4*	*2.8*	...	...
Pottery and related products	2.3	...	...	...	...
Farm machinery	...	...	...	...	6.0
Glass and glass products	*6.1*	...	...	...	...
Motor vehicles and motor vehicle equipment	...	*3.7*	*3.9*	*14.3*	*3.2*
Office, computing, and accounting machinery	...	...	...	3.7	...

a. ... indicates location quotient less than 2.0.

b. Italic figures indicate at least 1.0 percent of local work force in industry.

Table 10-7. Classification of 1900 Manufacturing Specialties by 1960 Location Quotient, for Twelve Established Centers of Manufactures (S indicates 1960 quotient 2.0 or more; L indicates 1960 quotient under 2.0)

Specialty	Bo	NY	Pa	Ba	Ci	SL	Cg	Pi	Bu	Cl	Dt	Ml
First Stage												
Tobacco manufactures	—	L	L	L	L	L	—	L	—	—	L	—
Canning and preserving	—	—	—	L	—	—	—	L	—	—	—	—
Meat products	—	—	—	—	—	—	L	—	—	—	—	—
Miscellaneous stone products	—	—	—	—	—	—	—	L	—	—	—	—
Planing mills	—	—	—	—	—	—	—	—	L	—	—	L
Beverage industries	—	—	—	—	S	S	—	—	—	—	—	S
Second Stage, Final												
Bakery products	—	—	—	—	L	—	—	—	—	—	—	—
Furniture and fixtures	—	—	—	—	L	L	L	—	L	—	L	L
Apparel and accessories	—	S	L	L	L	L	L	—	—	L	L	L
Knitting mills	—	—	S	—	—	—	—	—	—	—	—	L
Floor coverings	—	—	S	—	—	—	—	—	—	—	—	—
Indirect, Final												
Miscellaneous manufacturing	L	S	L	L	S	—	L	—	—	—	L	—
Footwear, except rubber	S	—	—	—	L	S	—	—	—	—	—	L
Other Categories												
Newspapers	L	L	—	—	L	—	—	—	—	—	—	L
Yarn mills	—	—	L	—	—	—	—	—	—	—	—	—
Paints and varnishes	—	—	—	—	—	—	—	L	—	—	—	—
Cutlery and hand tools	—	—	—	—	—	—	—	—	L	—	—	—
Drugs and medicines	—	—	—	—	—	—	—	—	—	L	—	—
Other printing and publishing	L	S	L	L	S	L	S	—	L	L	L	L
Other iron and steel*	—	—	L	—	L	—	L	S	L	S	L	S
Fabricated metal products	—	—	—	L	L	—	—	—	—	S	S	L
RR and misc. transportation equipment	—	—	—	—	L	S	S	—	—	—	—	L
Electrical machinery	S	—	—	—	—	—	S	L	—	L	—	—
Leather	—	—	S	—	—	—	—	—	—	—	—	S
Farm machinery	—	—	—	—	—	—	S	—	—	—	—	S
Blast furnaces	—	—	—	—	—	—	—	S	—	S	—	—
Structural metal products	—	—	—	—	—	—	—	S	—	—	—	—
Pottery and related products	—	—	—	—	—	—	—	S	—	—	—	—
Glass and products	—	—	—	—	—	—	—	S	—	—	—	—
Chemicals and allied products	—	—	—	—	—	—	—	—	S	—	—	—

* Included title "Foundry and machine shop products" in 1900; losses may be artifactual in some centers.

Note: Specialties are industries with location quotient of 2.0 or more which employed at least 1.0 percent of the local work force in 1900. Caption symbols are identified in Table 10-3.

of manufacture is new on the national scene, such as the aircraft specialties of Baltimore, Cincinnati, and St. Louis.

Detailed profiles in the manufacturing sector as of 1900 were compiled not only for the twelve centers near the northeastern manufacturing belt, but also for Washington and the distant regional metropolises of New Orleans, San Francisco, and Minneapolis. Washington had in 1900 no noteworthy specialties in the manufacturing sector, and commercial printing was its sole 1960 specialty. The regional metropolises had relatively few specialties in either year, but such specialties as were present seemed to represent an outgrowth of the center's regional orientation (Table 10-8).

None of the losses recorded in the regional metropolises fell within the first-stage categories. New Orleans retained its miscellaneous-food specialty of sugar refining; San Francisco retained its specialty in the canning and preserving of fruits and vegetables; grain milling remained a specialty of Minneapolis. Lost were five of the six other local specialties of 1900, including: apparel in New Orleans and San Francisco; footwear in Minneapolis; newspapers in Minneapolis; and other printing and publishing in San Francisco. Remaining was only an other-printing-and-publishing specialty in Minneapolis. Gained were a relatively few lines of nontraditional manufacture. As a consequence, the distinctiveness of the regional-center profile was no less in 1960 than in 1900.

Evidence on change in the structure of manufactures in Kansas City and Los Angeles, the newest of the established centers, is scant. Kansas City has had since the turn of the century a first-stage specialty, meat packing; but such ties as Los Angeles manufacturers then had with the southern California resource region disappeared. Among the present specialties of each center is a line of manufacture which is unambiguously new: motor vehicles in Kansas City; and aircraft in Los Angeles. Neither place yet matches older centers near the manufacturing belt with respect to the prominence of manufactures in the local economy or the diversity of local manufacturing specialties.

Our impression based on examination of the 1900 profiles in the manufacturing sector was that centers which recently had gained national prominence had relatively more specialties in new lines of industry than did older centers. This led us to think that new

Table 10-8. Industrial Profiles in the Manufacturing Sector for New Orleans, San Francisco, Washington, Minneapolis, Kansas City, and Los Angeles, 1960

Stage of Resource-Use, Type of Market, and Detailed Industry	LOCATION QUOTIENT FOR					
	NO	SF	Wa	Mp	KC	LA
First Stage, Nonfinal						
Misc. nonmetallic mineral and stone products	...[a]	...	...	6.3[b]	3.0	...
Petroleum refining	2.1	2.6	...	...	...	...
Grain-mill products	...	...	...	3.7	2.8	...
First Stage, Final						
Beverage industries	2.7	...	...	...	...	...
Misc. food preparations and kindred products	4.5	2.2	...	2.1	...	...
Canning and preserving fruits, vegetables, and seafoods	...	2.5	...	...	...	...
Meat products	...	...	...	2.4	2.9	...
Second Stage, Nonfinal						
Paints, varnishes, and related products	...	2.3	...	...	3.0	...
Misc. petroleum and coal products	6.8	...	...	2.4	3.2	...
Indirect, Nonfinal						
Printing, publishing, and allied industries, except newspapers	...	...	2.2	2.1	2.1	...
Electrical machinery, equipment, and supplies	...	...	...	...	...	1.7
Professional equipment and supplies	...	...	...	6.3	...	...
Misc. fabricated metal products	...	...	...	...	1.8	1.7
Misc. paper and pulp products	...	...	...	...	2.0	...
Fabricated structural metal products	...	...	...	...	2.0	...
Ship and boat building and repairing	3.6	4.3	...	...	...	...
Pottery and related products	...	...	...	...	...	2.1
Farm machinery	...	...	...	2.9	...	...
Motor vehicles and motor vehicle equipment	...	...	...	...	1.7	...
Office, computing, and accounting machinery	...	...	...	2.7	...	...
Indirect, Final						
Leather products, except footwear	...	...	...	...	2.2	...
Aircraft and parts	...	...	...	...	...	5.9

a. ... indicates location quotient less than 2.0.

b. Italic figures indicate at least 1.0 percent of local work force in industry.

centers broke into the ranks of the established centers by capturing a disproportionate share of industries new to and gaining importance on the national scene, rather than by capturing what once had been industrial specialties of the established centers. Our review of changes in the structure of manufactures in the leading centers of the pre-World War II period highlights the long-run persistence of local specialties or the continuity in the character of local manufactures. Equally important to our thesis, however, is the "newness" of specialties to be found in the centers which have gained prominence in the 1940s and 1950s. We preface our discussion with the caution that since 1900 growth of the work force in the manufacturing sector has proceeded relatively slowly on the national level and that, therefore, specialties of the new centers are likely to be found outside the manufacturing sector.

MANUFACTURES
IN THE NEW CENTERS

Of the four centers that shifted ahead of New Orleans in rank with respect to population size during the 1940s and 1950s, none could be said to specialize in manufactures in 1960. The proportion of the Seattle work force employed in the manufacturing sector was no higher than the corresponding proportion for the nation as a whole, and in Houston, Dallas, and Miami, the location quotient for all manufactures was less than unity. The manufacturing sector also was underrepresented in the industrial profiles of Atlanta, Denver, and San Diego, centers which now challenge New Orleans in the rank order with respect to population size (see Table 10-9).

In character of local manufactures, the new centers are a diverse group. Miami is distinguished by a dearth of manufactures in each resource-market category which is matched only by Washington among the established centers. Houston has as its foremost specialty the first-stage resource users producing for the non-final market and, in this regard, differs from any established center. Dallas, Atlanta, and San Diego resemble the older center of Los Angeles in that their only specialty was the "indirect, final" category of manufactures. Both the "indirect, final" and "first-stage, non-final" categories appear as specializations in Seattle, a combination not observed

Table 10-9. Location Quotients in Six Manufacturing Industry Categories, and Percentages of the Employed in Extractive and Construction Industries, for the New Centers, 1960 and 1950

Center	Symbol	Manufacturing, All	FIRST STAGE Non-final	FIRST STAGE Final	SECOND STAGE Non-final	SECOND STAGE Final	INDIRECT Non-final	INDIRECT Final	PERCENT IN— Extraction	PERCENT IN— Construction
						1960				
Houston	Ho	0.81	2.14	0.67	1.10	0.39	0.75	0.15	3.92	7.87
Dallas	Dl	0.82	0.48	0.92	0.41	1.08	0.85	1.33	3.86	7.67
Seattle	Se	1.02	1.40	0.72	0.46	0.49	0.57	5.67	2.21	6.28
Miami	Mi	0.44	0.15	0.68	0.24	0.97	0.39	0.41	2.62	7.29
Atlanta*	At	0.82	0.39	0.99	0.66	1.11	0.68	1.78	1.23	7.04
Denver*	De	0.70	0.49	1.06	0.33	0.46	0.88	0.72	3.02	7.27
San Diego*	SD	0.85	0.14	0.50	0.14	0.33	1.10	2.77	3.31	8.54
						1950				
Houston	Ho	0.86	3.23	0.69	0.82	0.48	0.73	0.15	3.71	10.31
Dallas	Dl	0.72	0.49	0.97	0.34	1.21	0.65	1.34	2.84	9.64
Seattle	Se	0.79	1.10	0.84	0.34	0.50	0.57	3.18	3.13	7.25
Miami	Mi	0.30	0.22	0.53	0.13	0.71	0.27	0.23	3.02	10.72
Atlanta*	At	0.75	0.52	0.92	0.95	1.16	0.61	0.43	2.05	8.08
Denver*	De	0.64	0.58	1.33	0.34	0.53	0.73	0.59	3.60	8.41
San Diego*	SD	0.54	0.14	1.08	0.09	0.31	0.29	3.61	7.92	8.21

* Had not displaced New Orleans in rank with respect to population size as late as 1960.

among the established centers. Finally, Denver with a disproportionate share of its work force in lines of manufacture classified as
first-stage users producing for the final market resembles such
regional metropolises as New Orleans, San Francisco, Minneapolis,
and Kansas City.

The strength of manufactures in the economy of the new
centers does not seem to be masked by an overrepresentation of
extractive and construction activity generated by their recent
growth. The proportions of the local work force employed in extraction and construction in the new centers in 1960 averaged 2.9
and 7.4, respectively. By way of comparison, the mean proportions
engaged in extraction and construction were 1.3 and 5.5 percent,
respectively, in the older centers. Although extraction and construction are relatively more prominent in the industrial structures
of the new centers, the difference is not sufficient to distort seriously
any comparison between groups or between individual centers in
the respective groups. (Illustratively, the maximum difference between any pair of the twenty-five centers with respect to the
proportion in extraction and construction is 6.7 percentage points.
Location quotients with a value Q in the center where extraction
and construction were at a minimum would equal location quotients with a value of $.933Q$ in the center where extraction and construction were at a maximum if the differential in the proportion
engaged in extraction and construction were eliminated.)

Examination of the detailed industrial profiles in the manufacturing sector for the new centers reveals that local specialties
are few in number (Table 10-10). The mean number of detailed
manufacturing industries for which the 1960 location quotient
took on a value of 2.0 or more, for example, stood at 3.3 in the
seven new centers. By comparison, the number of manufacturing
specialties averaged 8.6 in the twelve established centers with a
long-standing specialization in the manufacturing sector and 5.1 in
the remaining six established centers.

First-stage users appear as manufacturing specialties in the
leading centers so infrequently in 1960 that no case can be made
for their distinctiveness in the new centers. Sawmills and planing
mills had been present in Seattle, and prominent in the profiles
of some of the nation's established centers in 1900; they remained
a Seattle specialty in 1960, but had receded in importance in the

Table 10-10. Industrial Profiles in the Manufacturing Sector for Houston, Dallas, Seattle, Miami, Atlanta, Denver, and San Diego, 1960

Stage of Resource-Use, Type of Market, and Detailed Industry	LOCATION QUOTIENT FOR						
	Ho	Dl	Se	Mi	At	De	SD
First Stage, Nonfinal							
Petroleum refining	8.9[b]	...[a]	...	...	...	...	...
Sawmills, planing mills, and millwork	...	...	2.8	...	...	...	...
Structural clay products	...	...	...	...	...	2.1	...
First Stage, Final							
Misc. food preparations and kindred products	...	...	...	...	2.1	...	...
Second Stage, Nonfinal							
Misc. petroleum and coal products	...	2.4	...	...	...	...	...
Misc. chemicals and allied products	2.4	...	...	...	...	...	...
Second Stage, Final							
Misc. fabricated textiles	...	...	...	...	2.3	...	...
Indirect, Nonfinal							
Misc. machinery	2.0	...	...	...	...	...	...
Rubber products	...	...	...	...	...	3.9	...
Misc. fabricated metal products	...	...	...	...	...	2.8	7.5
Misc. paper and pulp products	...	...	...	...	2.1	...	...
Fabricated structural metal products	2.3	...	...	2.1	...	...	...
Ship and boat building and repairing	...	...	2.4	...	...	...	2.1
Railroad and misc. transportation equipment	...	...	2.6	...	...	...	...
Pottery and related products	...	...	...	...	...	2.9	...
Motor vehicles and motor vehicle equipment	...	...	...	...	1.6	...	...
Indirect, Final							
Leather products, except footwear	...	...	...	...	...	6.0	..
Aircraft and parts	...	2.4	12.0	...	2.7	...	5.8

a. ... indicates location quotient less than 2.0.

b. Italic figures indicate at least 1.0 percent of local work force in industry.

profiles of the older centers. Petroleum refining was gained as a specialty after 1900 by a new center, Houston, and by an established center, San Francisco, and became increasingly important in the economy of the old center of Philadelphia. No other first-stage

specialties appear in the profiles of the new centers, and few are
to be found in the profiles of the older centers. An "abrasive-
products" specialty had emerged in Minneapolis by 1960. Meat
packing had been lost as a specialty by Chicago between 1900
and 1960, but it had been gained as a specialty by Minneapolis and
St. Louis and remained prominent in the Kansas City profile.
The beverage industries had remained a specialty in Cincinnati,
St. Louis, and Milwaukee where they had been salient in 1900,
and sugar refining still was to be found in New Orleans.

Among second-stage users and industries using resources indirectly, the distinctiveness of the new centers with respect to
manufacturing specialties is clear-cut. We have enumerated the
industries that were both manufacturing specialties (location quotient of 2.0 or more) and major employers (at least 1.0 percent of
the local work force) in 1960 in the new centers along with the
industries which became manufacturing specialties between 1900
and 1960 and were by 1960 major employers in the established
centers (Table 10-11). Specialties are identified by the title of the
component industry employing the largest number of workers to
lend concreteness, insofar as the *Census of Manufactures* and
County Business Patterns permit a determination. On a detailed
industry basis, there is little overlap between this list of specialties
and the list of 1900 specialties. A scanning of the specialties list
suggests that specialties of the new centers, as well as the specialties
gained by the established centers since the turn of the century, typically are in lines of manufacture that developed after 1900.

It is reassuring on this point to examine the profiles of San Bernardino, San Jose, and Phoenix, centers that now are rapidly shifting
upward in rank with respect to population size. Among industries
that are second-stage resource users or use resources indirectly, the
highest location quotients occur in industries that would be identified by the titles: "cement, hydraulic," "blast furnaces and steel
mills," and "aircraft parts and equipment" in San Bernardino;
"ordnance," "computing and related machines," and "electronic
components" in San Jose; and "computing and related machines,"
"ready mixed concrete," and "aircraft parts and equipment" in
Phoenix.

Table 10-11. New Centers in Which a Given Industry Was Both a Local Specialty and a Major Local Employer in 1960, and Established Centers in Which the Industry Became a Local Specialty Between 1900 and 1960 and Was a Major Local Employer in 1960

Title of Component Industry Employing Largest Number of Workers	Center (cell symbols)
Aircraft engines, parts, and equipment	Ba Ci SD
Ordnance	De SD
Rubber products	De
Aircraft	SL LA Dl Se At
Fabricated structural metal products	KC Ho Mi
Oilfield machinery and equipment	SL Ho
Basic chemicals	SL Ho
Commercial printing	Wa KC Ho
Mechanical measuring devices	Ml Mp
Electrical industrial apparatus	Ml
Construction machinery and equipment	Ml
Motor vehicles and equipment	Bu Cl Dt Ml
Metalworking machinery	Cl Dt
Primary nonferrous industries	Pi Bu Cl
Blast furnaces and steel mills	Ba Cg Bu
Misc. fabricated metal products	Cg
Cleaning and toilet goods	Ci
Ship and boat building, repair	Bo Pa NO SF

Note: Cell symbols are identified in Tables 10-3 and 10-9.

177

Either a proposal that each center's industrial structure is but a replica of the national structure or a proposal that each center's industrial structure mirrors its own growth history might be met with skepticism. In working through the materials on manufactures in the leading centers, the localization of new lines of manufacture has seemed more salient than the redistribution of traditional lines of manufacture in generating change in the system of major centers. Insofar as this impression is accurate, a center's industrial structure should mirror its growth history.

The detailed manufacturing industries must be assigned an "age" if the relation between a center's contemporary industrial structure and the timing of population growth in the center is to be explored systematically. The age classification of industries proposed here is crude and approximate although there is little doubt that the so-called old industries were prominent on the national scene some decades before the new industries appeared. (Grouped as old industries are: all detailed industries in the leather and textile categories; the detailed industries in the furniture, lumber, and wood products category; ship and boat building and repairing; and within the food category, meat and grain-mill products, confectionary and beverage industries, and miscellaneous food preparations. The new-industries group is made up of: office, computing, and accounting machines; electrical machinery; the motor-vehicle and aircraft industries; professional and photographic equipment and supplies; synthetic fibers; petroleum refining, and miscellaneous petroleum and coal products; and rubber and plastic products. At the national level, a little over 6 percent of the work force was employed in old industries in 1960; the new industries accounted for a little less than 7 percent of the work force. All other detailed industries are grouped and designated intermediate, and together they accounted for some 15 percent of the nation's work force in 1960.)

The share of the national employment in old, intermediate, and new lines of manufacture, respectively, concentrated in each major center would be identical if the industrial structure of each center were a replica of the national industrial structure. That this is not the case is easily documented. The major centers typically have a lesser share of the nation's so-called old manufacturing activity

than of the intermediate or new manufacturing activity. Their
shares of intermediate and new manufactures relative to their
share of the old manufacturing activity are highly variable, moreover. The relative dearth of new manufactures in New Orleans or
the overrepresentation of new manufactures in Detroit, Los Angeles,
or Seattle "fits" an impressionistic classification of centers by age.
The overrepresentation of new manufacturing activity in Boston
or the underrepresentation of new manufacturing activity in Miami,
however, seems counter to an age-linked industry structure.

A more systematic check on the possible association between a
center's industrial structure and its growth history consists of
selected comparisons between: a center's percentage shares of the
national manufacturing work force in each age-of-industry group
as of 1960; and the center's percentage shares of the national population increase in each of three time periods, i.e., before 1860, 1860
to 1910, and 1910 to 1960. (The shares are reported in Table 10-12.)
We have regressed each center's share of old manufacturing industries on both the center's share of new manufacturing industries and the center's share of the national growth through 1860. The
partial regression coefficients in standard form are found to be
.4 and .7 for the respective measures. We also have regressed each
center's share of intermediate manufacturing industries on both the
center's share of new manufacturing industries and the center's
share of the national growth in the years 1860 through 1910. Again
the partial regression coefficients in standard form are found to be
.4 and .7 for the respective measures. We take these results to mean
that there is, indeed, a reflection of a center's growth history in its
contemporary structure.

An emphasis on the location of activities new in the national
economy rather than on the redistribution of ongoing activities
seems appropriate for understanding shifts over time in the relative
sizes of established centers and the entry of new centers into the
ranks of the established centers. Our results are partial in that they
pertain only to the manufacturing sector, but they may offer some
insight into the workings of a system of cities.

Berry (1965: 411 ff.) has summarized recent findings on city-
size regularities. A stochastic growth process, which assumes "that
probabilities of percentage growth of cities are independent of their

Table 10-12. Percentage Shares of 1960 National Manufacturing Work Force by Age of Industry, and National Population Increment in Given Time Periods, for Major Centers

| Center | MANUFACTURES | | | POPULATION INCREMENT | | |
	Old	Intermediate	New	1860	1860–1910	1910–60
Boston	1.78	1.34	2.34	1.58	1.82	1.14
New York	7.38	11.12	9.57	5.32	8.88	8.92
Philadelphia	2.76	3.62	3.34	2.88	2.25	2.40
Baltimore	0.84	1.23	1.08	1.04	0.73	1.11
Cincinnati	0.48	0.82	0.82	0.84	0.54	0.56
St. Louis	1.19	1.44	1.65	0.91	1.25	1.18
Chicago	2.99	6.44	5.58	0.89	4.17	4.61
Pittsburgh	0.52	2.57	1.13	0.98	1.92	1.08
Buffalo	0.50	1.13	1.32	0.61	0.71	0.79
Cleveland	0.52	1.76	2.11	0.30	0.94	1.31
Detroit	0.74	2.39	6.84	0.44	0.79	3.64
Milwaukee	0.68	1.09	1.49	0.28	0.63	0.84
New Orleans	0.40	0.29	0.12	0.62	0.28	0.58
San Francisco	1.31	1.35	1.12	0.27	1.14	2.32
Minneapolis	0.76	0.89	0.87	0.13	0.96	1.00
Kansas City	0.44	0.65	0.52	0.14	0.63	0.71
Los Angeles	2.24	3.88	8.28	0.03	0.87	7.17
Houston	0.34	0.69	0.55	0.03	0.18	1.30
Dallas	0.36	0.54	0.77	0.09	0.40	0.94
Seattle	0.59	0.38	1.37	0.00	0.57	0.88
Miami	0.28	0.27	0.10	0.00	0.02	1.07

size, and that there is a steady rate of addition of cities to the system at that lower threshold size which cities must satisfy to qualify as members of the system" (p. 412), would be consistent with many observed distributions of cities by population size, especially in countries where "the urban society is old and complex and has been influenced by large numbers of forces in many ways, such that the patterning effects of any one of these forces are lost" (p. 414). If a city's industrial structure at any point in time is conceived as a record of its past successes in capturing lines of activity new on the national scene, independence between size and subsequent growth rate may seem more plausible than if the structure is conceived as evolving in a more or less determinate fashion.

Chapter 11

CHANGE IN THE
FINANCIAL NETWORK

With the creation of the Federal Reserve System in 1913 and its becoming operational in 1914, an effort was made to keep banking practices compatible whenever possible with the earlier organization of banking. Federal Reserve Bank and Federal Reserve Branch Bank cities were designated which would be headquarters and branches, respectively, of the 12 banking districts into which the United States was divided. Most cities designated as Federal Reserve Bank or Branch cities in 1914 had been reserve cities, and the others later became reserve cities. Not all reserve cities have Federal Reserve Banks or Branches located therein, however.

The distinctions among country-bank, reserve, and central-reserve statuses were maintained. The Federal Reserve System continued to require that banks in cities with reserve status hold a larger percentage of reserves against their notes and deposits (Board of Governors of the Federal Reserve System, 1938: 962) although many potential advantages of the status were eliminated (St. Louis soon opted to relinquish its central-reserve status. The two remaining central-reserve cities, Chicago and New York, were reclassified as reserve cities in 1962. New York banks had declined in their proportion of the nation's deposits, and the Congressional legislation increased the funds available for loans by eliminating the special reserve requirements which had distinguished central-reserve cities [see Robins and Terleckyj, with the collaboration of Scott, 1960: 76–79, 201].)

When the Federal System started operating in 1914, the places holding reserve status in 1910 were designated reserve cities. New York, Chicago, and St. Louis remained central-reserve cities. Atlanta (Georgia) and Richmond (Virginia) also became reserve cities since they had been selected to head two of the twelve Federal Reserve Districts into which the country was divided (Board of Governors of the Federal Reserve System, 1938: 962). Not only have a number of cities lost, and others gained reserve status since 1914; several cities have done both.

Seventeen places were added to the roster of reserve cities before the Federal Reserve System had been in operation for a decade. The Federal Reserve Board's decision that all Federal Reserve Branch Bank cities should hold reserve status added five places to the roster (Board of Governors of the Federal Reserve System, 1943: 401). Six places gained reserve status at the request of local banks. Six other places gained reserve status at the initiative of the Federal Reserve Board and on the basis of their size and importance; among them was Buffalo, the only established major center lacking reserve status at the time.

(The five Branch Bank cities included Jacksonville [Florida], El Paso [Texas], Little Rock [Arkansas], Helena [Montana], and Charlotte [North Carolina], each of which still retained its special status in the mid-1960s. Birmingham [Alabama], Nashville [Tennessee], and Tulsa [Oklahoma] requested and retained reserve status, but Charleston [South Carolina] and Chattanooga [Tennessee], had given up their newly gained status before 1940 and Ogden [Utah] did so in the 1940s. Of the six places becoming reserve cities at the Board's initiative, Buffalo, Memphis [Tennessee], and Toledo [Ohio] retained their special status; Oakland [California] relinquished its status before 1940, Grand Rapids [Michigan], and Peoria [Illinois] dropped from the roster in the 1940s.)

Only five places had terminated their reserve-city designations and reverted to country-bank status by 1930. If one were dramatically inclined, the loss of reserve status could, in some cases, be described as the final financial blow to communities which were once important but failed to keep up with their rivals. Most notable is the case of Albany which had been one of the original redemption centers under the National Bank Act of 1864. In other instances, termination of reserve status represents the bitter outcome of a struggle between neighboring cities for financial supremacy, a struggle which started before any one of them was a significant banking center. The termination of reserve-city status for Tacoma (Washington), for example, marked the first elimination among contenders in the northwest, the ultimate victor being Seattle. (Other places terminating reserve status in the 1920s were Chattanooga and Charleston and Muskogee [Oklahoma].) Oakland (California), the last city to give up reserve status before America's entry into World War II, was overshadowed by neighboring San Francisco where some of the nation's largest banks were headquartered. By 1950 virtually all of Oakland's banks had become branches of institutions based in San Francisco.

The period since World War II has been one in which the relinquishment of reserve status was far more frequent than the acquisition of reserve status. Seven relatively small centers scattered across the nation were declassified in the 1940s. Only a single city gained reserve status, and an odd one at that—National Stock Yards (Illinois), a very small political unit located in the St. Louis metropolitan area and the site of a bank serving cattle-country banks. Between 1951 and 1965 another eight relatively small centers clustered in mid-continent were eliminated from the roster. Only one place gained reserve status, namely, Miami, the only major center not having held such status earlier. These shifts in status reduced to forty-eight the number of reserve cities in 1965 (see Table 11-1).

The shifts to and from reserve status have had the effect of strengthening the relation between a city's size and its status. By 1940 all but twelve of the nation's "top fifty" cities held reserve status. The exceptions are instructive: six places within a few hundred miles of New York City; three places in Ohio—Youngstown, Dayton, and Akron—encircled by the reserve cities of Pittsburgh, Cincinnati, and Cleveland; Oakland which had been overshadowed by San Francisco; Miami which was soon to gain reserve status; and San Diego to the south of Los Angeles on the Pacific coast. There were, in addition, twenty-three relatively small reserve cities, most of which were in trans-Mississippi territory, fourteen of which were to relinquish their reserve status by 1965. All but twelve of the nation's top fifty cities again held reserve status in 1965. Among them now were some rapidly growing western cities: Oakland; San Diego; Long Beach, a neighbor of Los Angeles; and Phoenix (Arizona). The 1965 roster of reserve cities included only ten smaller places, however, of which seven qualified for reserve status by virtue of their designation as a Federal Reserve Bank or Branch Bank city.

At the risk of oversimplification, changes in reserve status over the last hundred years may be viewed as describing the spread of metropolitan organization westward across the continent. As the trans-Mississippi territory was settled, fledgling metropolitan centers emerged. Such places, large by regional standards, but small relative to metropolitan centers in older sections of the nation, opted for reserve status. Some did become key points in the financial network that developed; others were unable to establish themselves as regional financial centers and terminated their special banking position.

The reserve-status series do not reveal shifts in the relative positions of the nation's major centers in the financial network. Most of the major centers held reserve status when the Federal Reserve System was established; each has maintained its special status. To detect changes in the financial prominence of the major centers, we must rely primarily on the selection of their financial institutions as underwriters of municipal bond issues and correspondent banks. In recording these changes, however, we must keep in mind that all represent metropolitan success stories.

STATE AND LOCAL BONDS

A major instrument for borrowing funds necessary for capital expenditures is the issuance of bonds. They are sold by various types of organizations: the federal government and some of its agencies; industrial, commercial, and transportation enterprises; public utilities; and states, cities, and other local political units such as school districts, toll-road agencies, sewerage districts, and counties. The ultimate geographical distribution of the funds obtained in the bond market by, say, a multi-establishment industrial firm or an agency of the Federal government is not readily apparent since the

Table 11-1. Reserve Cities in the National Banking System, by City Size in 1940, for 1910, 1940, and 1965 (x indicates reserve-city status)

Region and Center	1910	1940	1965	Region and Center	1910	1940	1965
Occupying Rank 50 or Higher in 1940							
Atlantic Seaboard				*Transmontane East—cont.*			
Boston	x	x	x	Birmingham, Ala.	...	x	x
New York	x	x	x	Memphis, Tenn.	...	x	x
Philadelphia	x	x	x	Nashville, Tenn.	...	x	x
Baltimore	x	x	x	Toledo, Ohio	...	x	x
Washington	x	x	x				
Miami	...	...	x	*Mid-Continent*			
				St. Louis	x	x	x
Atlanta, Ga.	...	x	x	Minneapolis[a]	x	x	x
Richmond, Va.	...	x	x	Kansas City[b]	x	x	x
Jacksonville, Fla.	...	x	x	Houston	x	x	x
				Dallas	x	x	x
Transmontane East							
New Orleans	x	x	x	Ft. Worth, Texas	x	x	x
Cincinnati	x	x	x	Oklahoma City	x	x	x
Chicago	x	x	x	Omaha, Neb.[c]	x	x	x
Pittsburgh	x	x	x	San Antonio, Texas	x	x	x
Buffalo	...	x	x				
Cleveland	x	x	x	*Far West*			
Detroit	x	x	x	San Francisco	x	x	x
Milwaukee	x	x	x	Los Angeles	x	x	x
				Seattle	x	x	x
Columbus, Ohio	x	x	x				
Indianapolis, Ind.	x	x	x	Denver, Colo.	x	x	x
Louisville, Ky.	x	x	x	Portland, Ore.	x	x	x
Occupying Rank 51 or Lower in 1940							
Atlantic Seaboard				*Mid-Continent—cont.*			
Albany, N.Y.	x	...	...	Cedar Rapids, Iowa	x	x	...
Savannah, Ga.	x	x	...	Dubuque, Iowa	x	x	...
Charlotte, N.C.	...	x	x	Galveston, Texas	x	x	...
				Lincoln, Neb.	x	x	...
Transmontane East				St. Joseph, Mo.	x	x	...
Grand Rapids, Mich.	...	x	...	Sioux City, Iowa	x	x	...
Peoria, Ill.	...	x	...	Topeka, Kans.	x	x	...
National Stock				Waco, Texas	x	x	...
Yards, Ill.	...	...	x	Wichita, Kans.	x	x	...
Mid-Continent				Des Moines, Iowa	x	x	x
Muskogee, Okla.	x	...	...	El Paso, Texas	...	x	x

Table 11-1. (continued)

Region and Center	1910	1940	1965	Region and Center	1910	1940	1965
			Occupying Rank 51 or Lower in 1940				
Mid-Continent—cont.				Far West.—cont.			
Little Rock, Ark.	...	x	x	Ogden, Utah	...	x	...
Tulsa, Okla.	...	x	x	Portland, Ore.	x	x	x
				Pueblo, Colo.	x	x	x
Far West				Salt Lake City, Utah	x	x	x
Tacoma, Wash.	x	...	...	Helena, Mont.	...	x	x
Spokane, Wash.	x	x	...				

a. St. Paul, a reserve city at each date, not shown separately.

b. Kansas City, Kansas, a reserve city in 1910 and 1940, not shown separately.

c. South Omaha, a reserve city in 1910, annexed in 1915.

monies may be used in almost any part of the nation or even overseas. The destination of funds obtained by state or local political
units is known, however; and bonds of this type commonly are referred to as "municipals."

The financial middlemen or "underwriters" involved in the sale
of municipals need a complete grasp of the money market since
their profit depends upon the difference between the price paid for
the bonds and their resale price. Most municipals are purchased by
underwriters on a competitive basis, that is, the financial intermediaries bid against one another for the issue in terms of who will
provide the capital at the lowest interest cost to the borrower. The
bidding calls for considerable skill since the bid must be low enough
to win the bond issue, but not so low as to make the interest on the
bonds unattractive to the middleman's prospective customers. The
margin between the purchase and resale price of the bonds is fairly
narrow and any need to "move" the bonds by raising their interest
yield, that is, lowering the resale price to the public, can erode
much or all of the underwriter's profit.

Taylor's (1957: 160) description may be useful:

Borrowing by state and local governments and special authorities
through the issue of securities involves the transfer of funds from
individuals and institutional savers to governmental authorities. The
investment dealer is the middle-man who performs this function. He
purchases the issues, often on the basis of competitive bidding, and
then sells them to the public, his profit consisting of the spread
between the buying and the selling price. Strictly speaking, of

course, an underwriter merely guarantees the sale of securities. Nevertheless, we shall use the term "underwriting" to include the purchase and re-sale of securities by investment dealers, since it is commonly used in this context.

An important feature of municipals, which distinguishes them from bonds presently issued by the federal government and private enterprise, is that interest earned on them is exempt from federal taxes. As a consequence, municipals are of particular appeal to lenders who would pay income tax on the interest of other issues and are unattractive to lenders exempt from income tax, such as pension funds or nonprofit foundations. The market for municipal bonds is, therefore, not the same as for other types of bonds, a fact which may influence the abilities of various intermediaries to market them. Also unique to municipals is the fact that commercial banks are allowed to underwrite their general obligations, whereas they have been prohibited since 1933 from underwriting other types of securities (Robinson, 1960: 103).

Municipal bonds are important issues in the security market. During the decade after World War II, they "accounted for more than one-fifth of the gross volume of new securities publicly offered. If federal offerings are omitted from this total, the proportion accounted for by state and local governments has been about one-third of the volume of new cash offerings in the public security market" (Robinson, 1960: 19).

Excluding the holdings of government agencies, half of all state and local government securities were held by individuals in 1950; a third were in the hands of commercial banks; and about 10 percent were owned by insurance companies, including fire, casualty, and life (Robinson, 1960: 216). Actually commercial banks were the single most important group of holders since a good part of the bonds owned by individuals are administered by the trust departments of banks. Holdings fluctuate from year to year, of course, influenced by the tightness of money and the tax situation—both actual and anticipated; "No other market," notes Robinson (1960: 68), "seems to have experienced such pronounced and frequent changes in customers."

Both the buying and selling dimensions of bond underwriting are related to the nature of the city in which the underwriters are located. First, the value of the securities underwritten generally exceeds the investment dealer's free capital. Unless the underwriter is a commercial bank, funds must be borrowed to finance the bonds between the time they are bought from the governmental body and the time they are resold. Moreover, bonds normally are accompanied by deposits to show "good faith." Thus, the operations of an underwriter are limited by the bank credit available,

> which in turn is tied in with the financial resources of the community served by the bank. Individual dealers can pool their resources when the size of the issue is greater than any one of them can handle alone, but here the financial resources of the community limit the size of the operation if the syndicate is composed entirely of dealers in any one city or area. [Taylor, 1957: 160]

Commercial banks often are desired members of an underwriting syndicate because they can finance the other underwriters and frequently hold municipals in their own investment portfolios. They also can help supply the capital necessary if an issue moves slowly in the re-offering stage because of some temporary jolts in the money market (Robinson, 1960: 104–105). Nevertheless, even commercial banks are affected by their own credit sources since there are limits on the amount they could prudently invest in bonds of a given issue. The limits are more restrictive for small banks, and, in this regard, the bank's capital resources affect its participation.

The ability of underwriters to resell municipals also is influenced by the characteristics of the cities in which they operate. Financial middlemen must sell their bonds as rapidly as possible; accordingly, in deciding whether to participate in a new issue, they are influenced by the potential market they can tap. Personal contacts with local institutions and wealthy investors as well as with networks of other dealers give the underwriters in some cities an advantage over those located elsewhere. The marketing advantages enjoyed by some commercial banks because of their financial functions and their location in an area with a concentration of institutional and individual purchasers of bonds has been described by Robinson (1960: 105) in the following way:

> The great commercial banks have close ties with many potential customers: their own country correspondents, trust departments of

related banks, wealthy individuals, and other financial institutions. The general knowledge by banks of who has money for investment probably gives them a rather substantial advantage in the marketing of state and local government obligations.

Samples of the municipal bond issues listed in the *Commercial and Financial Chronicle* in 1940 and 1964, respectively, were drawn for comparison with the 1900 sample of issues described in Chapter 6. For 1940 and particularly 1964, the task of identifying the geographic location of underwriters became increasingly complex. First, underwriters have become more likely to have offices in several cities; and the location of the main office may fail to reflect fully the locus of the distributional efforts. Second, the formation of syndicates is an increasingly common form of bond underwriting. The several "houses" forming the syndicate are likely to be located in different cities, and although a single organization usually heads a large syndicate, the appropriate geographic assignment is not always clear. (A directory, *Security Dealers of North America*, eighth edition [New York: Standard and Poor's Corporation, 1964], proved a useful supplement in locating the main office of the purchasing dealer[s] in 1964.) The increasing complexity in the organization of underwriters which gives rise to the geographic coding problem is, of course, itself an indicator of change in the spatial aspect of the financial network.

Perhaps the most noteworthy feature of the time series on the location of underwriters (Table 11-2) is the progressive deterioration of Cincinnati's prominence. Wade (1959: 165) has identified Cincinnati as "the focus of banking activity in Ohio" in the early 1800s, and in 1900 underwriters of municipals were more likely to be found in Cincinnati than in any other major center. As late as 1914 Ohio cities preferred Cincinnati's designation as a Federal Reserve Bank city rather than Cleveland's. By 1940 Cincinnati's competitive position in the underwriting of municipals had deteriorated, however. Only 6 percent of the municipals were underwritten by Cincinnati firms, as compared with 10 percent for New York or Chicago firms. At the most recent date, 1964, Cincinnati firms served as underwriters for 4 percent of the municipal issues, as compared with 38 percent for New York firms and 22 percent for Chicago firms. In fact, thirteen of the major centers were found to have

Table 11-2. Percentage of Municipal Bond Issues Underwritten by Financial Intermediaries Headquartered in a Major Center in 1900, 1940, and 1964

Location of Underwriter	1900	1940	1964	Location of Underwriter	1900	1940	1964
Atlantic Seaboard				**Mid-Continent**			
Boston	4.3	3.5	11.9	St. Louis	0.7	2.5	7.6
New York	8.6	10.9	37.5	Minneapolis	1.3	3.9	9.4
Philadelphia	3.0	3.1	8.6	Kansas City	0.3	1.9	5.8
Baltimore	0.3	0.2	3.8	Houston	...	1.2	3.0
Washington	...	0.2	3.0	Dallas	...	1.0	5.6
Miami	...	0.4	1.3	**Far West**			
Transmontane East				San Francisco	0.7	1.6	7.1
New Orleans	0.3	1.0	3.0	Los Angeles	...	1.4	3.5
Cincinnati	14.3	5.6	3.8	Seattle	...	0.8	4.3
Chicago	9.6	10.1	22.5	**Other**			
Pittsburgh	0.3	2.7	4.8	All other cities	40.2	49.0	46.8
Buffalo	...	0.8	1.8	Government agencies	5.3	7.2	8.1
Cleveland	6.3	1.6	5.1	Unknown	6.3	4.1	6.8
Detroit	...	1.4	9.4				
Milwaukee	...	1.2	3.3				

concentrations of underwriters greater than the Cincinnati concentration.

Cincinnati's initial leadership had been based on its strong position within its own east-north-central region and in the contiguous east-south-central states (Table 11-3). Its lessening activity could, then, result either from a loss of leadership in its hinterland bond market or from a decrease in the proportion of municipal issues originating in the Cincinnati hinterland. In point of fact, its deterioration was an outcome of both factors.

The initially strong positions of New York and Chicago, on the other hand, have been consolidated. There is, as yet, no indication that the participation of these cities in the bond market has reached a peak; and no challenger to their preeminence has emerged. Between 1900 and 1940, neither city increased its share of business substantially but neither city lost ground as did Cincinnati. New York's strength remained concentrated primarily in the mid-Atlantic states, and Chicago continued to be especially active in the north

Table 11-3. Percentage of Municipal Bond Issues from Each Geographic Division Underwritten by Financial Intermediaries Headquartered in a Major Center in 1900, 1940, and 1964 (. . . indicates less than 10 percent underwritten)

Location of Underwriter, Center and Division	GEOGRAPHIC DIVISION AND YEAR								
	NEW ENGLAND			MIDDLE ATLANTIC			SOUTH ATLANTIC		
	1900	1940	1964	1900	1940	1964	1900	1940	1964
Boston (NE)	68.4	45.0	43.5	...	...	11.1	...	...	16.3
New York (MA)	...	15.0	47.8	29.7	40.0	69.8	...	14.6	44.2
Philadelphia (MA)	...	...	...	12.5	16.5	27.0	...	...	18.6
Pittsburgh (MA)	...	...	...	...	14.1	12.7	...	...	...
Baltimore (SA)	...	...	...	...	...	...	...	...	11.6
Cincinnati (ENC)	...	...	...	...	...	...	...	14.6	...
Chicago (ENC)	...	...	21.7	...	...	20.6	...	...	23.3
Detroit (ENC)	...	...	...	...	...	...	...	...	11.6
Lesser centers	10.5	40.0	34.8	42.2	24.7	54.0	63.6	78.0	65.1

	EAST-SOUTH-CENTRAL			EAST-NORTH-CENTRAL			WEST-NORTH-CENTRAL		
	1900	1940	1964	1900	1940	1964	1900	1940	1964
Boston (NE)	...	...	...	...	...	...	...	...	13.0
New York (MA)	...	...	26.8	...	...	37.2	...	...	20.4
Cincinnati (ENC)	33.3	...	...	35.7	20.0	...	...	...	...
Chicago (ENC)	16.7	...	...	...	26.4	44.9	25.6	17.4	11.1
Cleveland (ENC)	...	...	...	11.2	...	...	...	...	...
Detroit (ENC)	...	...	...	...	...	21.8	...	...	...
Milwaukee (ENC)	...	...	...	...	...	10.3	...	...	...
Minneapolis (WNC)	...	...	...	...	...	...	...	13.0	48.1
St. Louis (WSC)	...	...	...	...	...	10.3	...	...	...
Lesser Centers	41.7	85.7	61.0	37.8	33.0	48.7	27.9	56.5	24.1

	WEST-SOUTH-CENTRAL			MOUNTAIN			PACIFIC		
	1900	1940	1964	1900	1940	1964	1900	1940	1964
Boston (NE)	...	...	...	...	...	...	...	...	16.7
New York (MA)	10.0	...	22.5	...	16.2	...	...	...	31.0
Chicago (ENC)	10.0	...	20.0	33.3	...	18.2	...	...	14.3
Detroit (ENC)	...	...	...	...	...	...	...	...	14.3
New Orleans (WSC)	10.0	...	15.0	...	...	...	...	...	...
St. Louis (WSC)	...	...	10.0	...	...	...	...	...	...
Kansas City (WSC)	...	...	12.5	...	...	36.4	...	...	...
Houston (WSC)	...	10.9	17.5	...	...	...	...	...	...
Dallas (WSC)	...	...	25.0	...	...	...	...	...	...
San Francisco (P)	...	...	...	...	...	...	...	26.7	40.5
Los Angeles (P)	...	...	...	...	...	...	...	16.7	23.8
Seattle (P)	...	...	...	...	...	18.2	...	10.0	28.6
Lesser Centers	40.0	49.1	50.0	58.3	62.2	63.6	61.9	50.0	28.6

Note: Major centers, identified in Table 11-2, are shown only if they participated in 10 percent of issues at some date; "lesser centers" includes all places except major centers.

central region. Growth in the volume and territorial dispersion in their underwriting activity since 1940 has been striking, however. By 1964, the underwriting role of New York and, to a lesser extent, Chicago had expanded into nearly every section of the United States. Although a number of centers now are important within a proximate region, only New York and Chicago play a significant role throughout the nation.

Since for centers other than New York and Chicago underwriting activity takes place in a regional rather than national context, changes in regional prominence are perhaps most informative.

Boston has slowly given way to New York as the leading underwriting center for issues originating in New England, and Chicago has begun to play a more active role in recent decades.

In the mid-Atlantic states, New York has increased still further its considerable margin over Philadelphia. Philadelphia now is being challenged by Chicago, which has risen in importance since World War II.

None of the major centers captured a large share of the underwriting of issues originating in the south Atlantic states in 1900 although New York and Cincinnati were the most active. In recent decades, New York has expanded its activity; but Cincinnati has receded into a minor position. Chicago now is second only to New York in underwriting the region's issues.

Located on the northern edge of the east-south-central states, Cincinnati was the leading center for underwriters of bond issues originating in the region at the turn of the century. Cincinnati along with Chicago, the second-ranking center of 1900, has declined in importance. The activity of New York has increased since 1940, and by 1964 New York had become the leading center for the region's issues.

In 1900 Cincinnati also had been the leading center in its own region, the east-north-central states, with Cleveland occupying second rank. Chicago had overtaken these cities by 1940 and has expanded even further since World War II, however. It is now New York which threatens to become the prime factor in the region, rather than a center located within the region itself.

Chicago, the dominant center in the west-north-central states in 1900, declined as Minneapolis captured a steadily growing share of the issues originating in the region. New York's gains since 1940

now rank it second only to Minneapolis in underwriting bonds issued in the west-north-central states.

Competition for bond issues in the west-south-central states
comes from a number of sources. Both Chicago and New York have
extended their roles in the last few decades. The regional centers,
New Orleans, St. Louis, Kansas City, Houston, and Dallas, also play
an active role in financing their own region, however.

The pattern of change in the mountain states is mixed. Chicago
appears to have lost and regained importance in the region, and
New York's importance appears to have declined from a 1940 peak.
Kansas City is currently the most active of the major centers in
underwriting issues originating in the mountain states.

On the Pacific coast, San Francisco, Los Angeles, and Seattle
houses are important middlemen for the region's bonds; but New
York has been gaining significance since World War II.

A final observation is directed to the gains in underwriting activity registered by nearly all centers between 1940 and 1964. Increasingly through the years, issues are purchased by syndicates
which consist of several underwriters who may be headquartered in
different cities, rather than by a single bank or investment house.
There is a "built-in boost" for the cities if underwriters in a number
of cities participate in floating a given bond issue, for the issue will
be registered as "underwritten" in each of the several centers. An
indication that participation in syndicates occurs with relatively
high frequency among financial houses located in major centers is
the fact that no corresponding gain in activity is recorded for the
aggregate of lesser centers (Tables 11-2 and 11-3). The reorganization of underwriting practices seems to have worked to the advantage of established centers, and a few case studies help to
substantiate this point.

SYNDICATES IN THE MUNICIPAL
BOND MARKET

Many small bond issues still are underwritten by a single firm,
or a group of a few firms, located in the region from which the
bonds are issued. Nearly half the municipal bond issues in the samples selected for study had intermediaries located outside the major

centers, for example. Granted that in some instances these issues were underwritten by firms headquartered in cities such as Atlanta, Denver, and Louisville which would qualify as regional financial centers on a par with some of the so-called major centers, the fact remains that local institutions are of considerable importance for small issues. Taylor's survey of state and local bonds issued in the southeastern United States indicates that southern firms have underwritten the overwhelming majority of issues under 100,000 dollars and most of those up to 500,000 dollars. There are no great centers of capital in the region, however; and as a consequence, nonsouthern financial intermediaries are very active in underwriting the larger municipal issues of the area (Taylor, 1957: 162).

Many larger bond issues involve syndicates formed by several commercial and/or investment bankers, often located in different cities. These syndicates are likely to include institutions with differing skills that can contribute to the success of the venture. For example,

> some firms have "good distribution" (a large or effective sales organization) but they do not have the capital to act as major underwriters. Other firms stand in the opposite position of having ample capital to take underwriting risks but of having somewhat less broad distributive outlets. Still other firms may have the kind of aggressive executives that make them natural managers; others may lack such talent and leadership. Such specialization weighs in the formation of a syndicate. [Robinson, 1960: 123]

New York and Chicago stand out with respect to frequency of participation in these intercity groupings. Although data are not available on the value of the bonds apportioned to each city's participants in a syndicate, the vital roles of New York and Chicago can be determined on the basis of other evidence. Large syndicates are headed by one or two firms which organize the bids and the distribution of the bonds if the bid is successful. During the period sampled in 1964, New York City institutions headed alone or jointly twenty-six of the seventy-five such syndicates recorded; Chicago was second with seventeen; and Minneapolis was a distant third with four.

New York and Chicago firms are more likely than firms in other centers outside the region of issue to participate in intercity syndicates. For example, most cities located outside the west-south-central states were represented by only a single firm in each of the region's bond issues in which they participated during 1964. The two largest issues in the sample period were for 15 million dollars each. In one case the syndicate consisted of thirty-one organizations, with ten major centers involved. Two of the ten participating centers were located in the region of issue: Dallas represented by one firm; and New Orleans represented by ten firms. Six of the eight extraregional participating centers were represented by a single firm, namely: Boston; Washington; Cleveland; Detroit; St. Louis; and Kansas City. Seven New York underwriters and two Chicago underwriters participated in this issue originating outside their region, however. The second issue was floated by a syndicate consisting of thirty-two underwriters, including one from Minneapolis and St. Louis, respectively, two Kansas City firms, four Chicago institutions, and ten from New York City.

Most centers participate outside their own region only occasionally when there is a major bond issue. These same cities may partipate both more frequently and with greater intensity in the syndicates formed for bonds issued in their own region, but that is as far as they go. By contrast, New York and Chicago are active in most intercity syndicates that are formed; and often several of their underwriters are in the group. Consider a 100 million-dollar bond issue from the Pacific area during 1964. The winning syndicate included 144 different houses headquartered in more than twenty different cities, including most of the major centers (Table 11-4). There were thirty-seven New York firms and twelve Chicago firms involved. The only other center with as many as ten firms participating was San Francisco, a center within the region, represented by sixteen firms. Consider the participation in the next two largest Pacific flotations during the same sample period, each a 15 million dollar issue. The syndicate consisted of fifty-six underwriters in one case, and thirty-seven underwriters in the other. Again New York provided the largest number of participating firms, twenty and thirteen, respectively.

The total number of Pacific coast issues in which at least one underwriter in each center participated during the sample period is of particular interest (Table 11-4). The three large bond issues account for the total participation of nearly half the centers. San

Table 11-4. Number of Underwriters from Each Major Center Participating in Three Largest Pacific Municipal Bond Issues, and Number of Pacific Issues Underwritten in Each Major Center, for a 1964 Sample Period

Location of Underwriter	NUMBER OF UNDER-WRITERS BY SIZE OF ISSUE (MILLIONS)			NUMBER OF ISSUES BY SIZE (MILLIONS)		
	$100	$15	$15	All Issues	Under $15	$15 or more
Atlantic Seaboard						
Boston	4	1	4	7	4	3
New York	37	20	13	13	10	3
Philadelphia	2	2	0	3	1	2
Baltimore	2	1	1	4	1	3
Washington	1	1	0	2	0	2
Miami	1	0	0	1	0	1
Transmontane East						
New Orleans	0	0	0	0	0	0
Cincinnati	3	0	0	1	0	1
Chicago	12	4	2	6	3	3
Pittsburgh	2	1	1	3	0	3
Buffalo	0	0	0	1	1	0
Cleveland	7	0	2	3	1	2
Detroit	1	1	1	6	3	3
Milwaukee	2	0	0	1	0	1
Mid-Continent						
St. Louis	4	1	0	2	0	2
Minneapolis	2	0	0	3	2	1
Kansas City	0	0	1	3	2	1
Houston	1	0	1	2	0	2
Dallas	4	3	0	2	0	2
Far West						
San Francisco	16	9	2	17	14	3
Los Angeles	5	2	2	10	7	3
Seattle	3	1	3	12	9	3
Other						
Cities and unknown	35	9	4	15	12	3
Government agencies	0	0	0	3	3	0

Francisco, Los Angeles, and Seattle in the far west participated in a sizable number of smaller Pacific bond issues as well as in the three issues of $15 million or more. New York was the only center outside the Pacific area which equalled these regional centers in the underwriting of both large and small Pacific issues, however.

Another way of thinking about these results is that if one were to select a major offering from any region, the chances are high that New York City and Chicago firms would be prominent and that some major cities in the region would be particularly active. The participation of underwriters headquartered in centers outside the region probably would be minimal, aside from New York or Chicago firms. New Orleans, for example, was very active in one of its region's major offerings, but did not participate in any Pacific issue.

The regional patterning is a result of the marketing factor, that is, there is frequently a better market for municipal bonds within the region of issue. Hence, local investment houses and banks may have an advantage in selling to the potential purchasers (Robinson, 1960: 106). The pattern is reinforced by certain kinds of pressures that develop, such as that on local banks to support the issues of their communities. The participation of New York and Chicago in many of these larger syndicates has the effect of linking the various regional underwriters into the national money market.

Chapter 12

THE SELECTION OF CORRESPONDENT BANKS

Although the American economy is based on a free-flowing interchange between all parts of the nation, banks in the United States are either locally based independent units or parts of chains which cover, at most, a single state. A smooth financial flow between areas is vital, however, since most forms of economic and social interaction are based on monetary transactions (Lieberson and Schwirian, 1962: 69). In some countries, such as Great Britain and Canada, there are nationwide bank chains which allow a check written on a bank in one area to be cashed in a branch in a distant region. By contrast, a check written by, say, a shoe wholesaler on his California bank for shoes sent from a St. Louis manufacturer is deposited in a bank which is completely independent of the bank on which the check was written. Nonetheless, the check must be fully negotiable and credited to the depositor quickly.

Although many checks now are cleared through Federal Reserve Banks, smaller banks in particular still rely on their correspondent banks located elsewhere in the nation to help clear checks received on out-of-town banks (Lieberson and Schwirian, 1962: 72–73). Correspondents also facilitate the transfer of funds required by an outlying bank's customers as well as

... serve a multitude of special bank needs, such as the purchase and holding of securities on behalf of banks located away from the major financial markets [or] supply[ing] credit information about prospective borrowers and aid in making loans, for example, when a borrower requires a loan of a size which exceeds a bank's legal or prudent limit. [Lieberson and Schwirian, 1962: 73]

Other functions of correspondents, according to Finney (1958: 10–17), include: analyzing the investment portfolios of smaller banks; providing technical help in connection with unusual or new types of loans; shipping currency; handling foreign exchange; advising on bank procedures and public relations; providing general investment services; keeping banks informed about relevant developments in finance and government; acting as depositaries of reserves for commercial banks that are not members of the Federal Reserve System; and helping the smaller banks gain new business when a major customer opens operations in their community.

In brief, lesser banks receive services from their correspondents which are based on the latter's advantages of size and location. Very large banks can maintain financial specialists and offer services of a highly technical nature which would prove uneconomical for smaller banks. Correspondents in Federal Reserve Bank or Branch cities or in major money markets can provide special services which hinge on their proximity to certain activities. Finney (1958: 19) notes that "if our system were one of nationwide branch banking many services currently rendered by the correspondent would likely be performed by the head office."

Competition among large banks for correspondent business is intense. Banks maintain balances on deposit in their correspondent banks; and these balances, which do not pay interest, may be used by the correspondent for loans and other profitmaking ventures. Although declining in importance, bankers' balances amounted to 7 percent of all deposits in commercial banks in 1954 (Finney, 1958: 53). In the same year, what were then the central-reserve city banks of New York and Chicago held more than 4.5 billion dollars of noninterest bearing commercial bank deposits (Finney, 1958: 30). Madden (1959: 17) mentions one New York bank which alone held 1.6 billion dollars in deposits from other banks at the end of 1958. To give one an idea of how beneficial such interest-free deposits might be, consider that 6 percent interest on these funds for one year would yield 9.6 million dollars. Other advantages accrue to the depositary bank as well. Banking connections established in the communities in which the smaller banks are located may result in new customers for the depositary bank. Fees are to be gained through security sales and other transactions performed for the smaller banks.

MAJOR CENTERS AS CORRESPONDENT-BANK LOCATIONS

Samples of banks were drawn from the 1940 and 1965 listings in the *Rand McNally Bankers Directory* for comparison with the sample drawn from the 1900 listing. Again each bank was classified by the geographic division in which it was located, and each major center in which the bank had a principal correspondent was recorded. The proportions of banks selecting a correspondent in a given center at the three dates offer a basis for assessing change in the prominence of that center in the financial network.

New York's correspondent position has weakened since 1900 although it still ranks first among the major centers as a choice for the location of a correspondent bank. Its prime position in 1965 was based on its selection by less than two-fifths of all banks in the nation and about half the "independent" banks, that is, banks which were not branches in a banking chain. At the turn of the century, on the other hand, nearly four-fifths of the nation's banks, which then included few branches, had selected New York. Most of the decline in New York's attractiveness appears to have occurred before World War II (see Table 12-1).

Chicago has remained in second place as a correspondent-bank center. Although a decrease in the proportion of the nation's banks served is recorded, the decline in Chicago's correspondent position is much less sharp than that observed for New York. The three cities that vied for third place in 1900 also have lost ground. The frequency with which Boston served as a correspondent location dropped sharply between 1900 and 1940, and Philadelphia's decline has been particularly severe since 1940. St. Louis has been able to maintain a substantially stronger position than its early rivals, but there is no suggestion of expansion in correspondent services.

Table 12-1. Percentage of Banks Selecting as a Principal Correspondent a Bank Located in a Major Center in 1900, 1940, and 1965

Location of Correspondent	INCLUDING BRANCHES			EXCLUDING BRANCHES	
	1900	1940	1965	1940	1965
Atlantic Seaboard					
Boston	9.8	5.9	4.5	6.5	6.0
New York	77.9	49.2	37.7	53.9	50.1
Philadelphia	9.2	8.8	4.3	9.7	5.7
Baltimore	2.3	1.5	2.2	1.7	2.7
Washington	0.2	1.2	0.6	1.3	0.8
Miami	…	0.7	1.2	0.8	1.6
Atlanta, Ga.	0.4	3.5	4.7	1.8	6.3
Transmontane East					
New Orleans	4.0	2.0	2.2	2.1	3.0
Cincinnati	5.8	3.9	1.8	4.4	2.2
Chicago	35.2	25.3	24.0	27.7	32.2
Pittsburgh	2.5	4.7	3.3	5.1	3.8
Buffalo	…	…	1.0	…	0.5
Cleveland	2.7	2.7	2.0	2.8	2.7
Transmontane East—cont.					
Detroit	2.3	1.7	2.9	1.9	3.8
Milwaukee	2.1	4.9	3.7	5.3	4.9
Mid-Continent					
St. Louis	11.5	8.6	8.4	9.5	11.2
Minneapolis	5.8	7.6	8.8	8.3	11.7
Kansas City	5.8	9.3	8.6	10.4	11.2
Houston	0.8	2.4	3.3	2.7	4.4
Dallas	0.8	2.7	7.3	3.0	9.8
Far West					
San Francisco	1.0	5.1	8.1	3.0	3.3
Los Angeles	0.6	2.4	5.1	1.1	3.0
Seattle	…	0.7	1.4	0.6	1.1
Other					
All	37.3	60.0	65.8	61.5	69.2

Note: The number of branch banks increased from 119 in 1900 to 3,500 in 1940 and 12,000 in 1965. Branches located in the same community as the main office of their chain are excluded.

Deterioration in the positions of these established centers of finance has occurred during a period in which the average number of different correspondent cities selected by the nation's banks increased from 2.4 to 3.3. (Correspondent selections in cities other than the twenty-three places identified in Table 12-1 have been counted as one even if more than one nonspecified city was selected by a given bank.) One reason for the decline in strength, if not rank of these established centers is the rise of urban competitors that in 1900 were of little consequence. Kansas City and Minneapolis have expanded their correspondent services and, by 1965, were at least on a par with St. Louis as centers for correspondent-bank selections. In recent decades, Dallas has emerged as an active competitor of the older centers.

Another important characteristic of bank-correspondent services during this century is their diffusion over more cities. In 1900 fewer than two-fifths of all banks had correspondents in a city other than the major centers of 1960 or Atlanta (Georgia), which now outranks all places east of the Mississippi other than New York and Chicago as a correspondent center. By 1965, two-thirds did so. Only in part can this change be regarded as an artifact of the inclusion of places which were to undergo very marked growth after 1900 and become "major" centers by 1960. The change seems also to reflect the larger number of principal correspondents listed by banks which, in turn, is perhaps due to the expansion of intercity relations and interdependencies which extend beyond the major centers.

Finney's analysis of the proportions of interbank deposits held by institutions in various cities is relevant. The results tend to support the notion that correspondent services have become more widely diffused. She notes, for example, that between 1934 and 1954 the percentage of all interbank deposits held by New York central-reserve banks declined from 38 to 26 percent although the absolute amount more than doubled during the period. Moreover, New York's share had fallen to 15 or 16 percent before 1960. According to Finney (1958: 30), "This decline reflects the long-time trends of the general diffusion of industry throughout the country and the reduced importance of the nongovernmental securities markets."

What has happened is that the reserve-city banks have increased their positions as depositaries for the lesser "country" banks; for example:

> Interbank deposits have generally shifted away from New York City and the New England and Middle Atlantic regions toward the South, Southwest, and Middle West, and consequently they are now less unevenly spread throughout the districts than they were in 1934. . . . Banking developments have accompanied the industrial expansion of the newer regions of the country and improvement of the agricultural situation in the past twenty years. . . . Banks in the South and Southwest held increasing proportions of balances, rising from 8 to 18 per cent in the Dallas district and from 6 to 10 per cent in the Atlanta district. . . . Correspondingly, balances owned by banks in New England, the Middle Atlantic states and the Middle West declined in relative importance. [Finney, 1958: 39–44]

A more complex pattern of economic relationships has developed which has worked to the detriment of some important early financial centers. It has been a great boon to their rivals in the more rapidly growing parts of the nation, however. Finney (1958: 24–25) describes a process in which the country banks and the banks which are not members of the Federal Reserve system tend to concentrate their deposits in outlying reserve-city banks. The latter, in turn, maintain the bulk of their deposits in the major money centers such as New York and Chicago. In recent decades, the outlying centers and their subordinates have grown more rapidly than the national money market banks.

Another factor influencing differentially the growth of centers with respect to correspondent-banking activity has been the expansion of chain banking, particularly in the western states. The number of commercial bank branches increased from 119 in 1900 to 3,500 in 1940 and 12,000 in 1962 (*Historical Statistics of the United States*, Series X156). Since these branches normally use their headquarters for correspondent functions such as check clearings and other banking services, the development of banking centers in branch-bank states is facilitated. California, for example, has several statewide networks of banks headquartered in San Francisco, in Los Angeles, or in both centers. As a consequence, when the correspondent selections of branch banks are excluded from the samples for 1940 and 1965, the growth of correspondent-banking activity in San Francisco and Los Angeles is diminished. Banking centers in states where statewide chains are prohibited, of course, appear more active relative to centers such as San Francisco and Los Angeles when the chains are eliminated and only the correspondent selections of "independent" banks are considered.

Although the percentage of banks using correspondents in any of the centers located in states which prohibit chain banking could only rise were chain banks to disappear, it is not clear how much shifting in the relative positions of centers would occur. Many chain banks now served by headquarters in San Francisco or Los Angeles probably would have principal correspondents in these centers even if there were complete freedom of choice with respect to correspondent banks.

Although discussions about "what might be" are nothing more than crude speculation, it is noteworthy that the centers which headquarter the major statewide chains in California are the state's largest metropolitan centers. Hence, the most appropriate single set of data for measuring change over time in the location of correspondent banks would seem to be the series which includes branch-bank choices. An alternative series in which branch-bank choices are excluded, probably tends to overestimate the growth of correspondent activity in centers where chain banking is prohibited and to underestimate the increase in centers which serve as headquarters of bank chains.

Overall, one may be struck either by the continuing preeminence of New York and Chicago and the consistently minor importance of centers such as Washington and Buffalo or by the expansion of correspondent-bank services in such places as Minneapolis, Kansas City, and Dallas. A measure of association (*tau*) between the ranks of the twenty-three centers separately identified in 1900 and 1965 may serve as a useful guideline; the coefficient takes on the value of .42. The corresponding measures are found to be .67 between 1900 and 1940, a forty-year span, and .69 between 1940 and 1965, a span of 25 years. Some tendency toward stability in the spatial pattern of correspondent banking over this century is detectable, but there has been considerable flux, especially since World War II.

The hinterland of New York remains more nearly national in scope than that of any other major center. At each date New York was the leading correspondent center not only in its own division, the middle Atlantic states, but also in the south Atlantic, east-south-central, and mountain divisions; and New York has remained second only to Boston in New England. By 1940 it had been displaced from first to second rank by Chicago in the east-north-central states, from first to third rank by Chicago and Minneapolis in the west-north-central states, and from first to second rank by San Francisco in the Pacific states. Since 1940 Dallas has shifted ahead of New York in the west-south-central states, and Los Angeles has displaced New York from second to third rank along the Pacific. (The geographic patterning of hinterlands is shown in Table 12-2. Rankings based on these data refer only to position among the twenty-three centers identified separately in Table 12-1. A lesser center could have been a more frequent choice than any center identified separately.)

The only other center which might be said to have the nation for its hinterland is Chicago. Its claim is less secure than New York's claim, however. New York offers more effective competition to Chicago in the midwest, Chicago's stronghold, than Chicago can offer New York in the area of the latter's primacy. Moreover, Chicago is selected much less frequently than New York in those sections of the nation where another of the major centers has primacy. Nonetheless, no deterioration in Chicago's position vis-à-vis New York is evident; neither has another rival for the national hinterland appeared. Although St. Louis, Minneapolis, Kansas City, and Dallas are active correspondent-bank centers, each serves a rather restricted contiguous hinterland.

From a regional perspective, many older centers have declined to a fraction of their earlier strength; and new centers have been catapulted into important regional niches.

New England has been relatively stable, with Boston strengthening its supremacy while New York declined. A major change has been the growing proportion of New England banks which select a lesser center as the site of one of its principal correspondents. In the mid-Atlantic division, New York's dominance still is unchallenged; but Philadelphia's edge over Pittsburgh has dwindled. Lesser centers have remained relatively unimportant as correspondent-bank sites for banks in the middle Atlantic states.

A new banking center has been created within the south Atlantic states at the expense of older centers located outside the area. New York, Philadelphia, and Baltimore lost ground during the period that Atlanta rose from a minor position to the second rank as corre-

Table 12-2. Percentage of Banks in Each Geographic Division Selecting as a Principal Correspondent a Bank Located in a Major Center in 1900, 1940, and 1965 (... indicates less than 10 percent selecting)

Location of Correspondent, Center and Division	GEOGRAPHIC DIVISION AND YEAR								
	NEW ENGLAND			MIDDLE ATLANTIC			SOUTH ATLANTIC		
	1900	1940	1965	1900	1940	1965	1900	1940	1965
Boston (NE)	69.2	83.9	74.0	12.7	...	...	...	...	...
New York (MA)	61.5	72.6	46.0	94.4	87.5	64.0	84.7	47.8	55.7
Philadelphia (MA)	...	...	...	49.3	45.5	28.0	16.9	14.9	...
Pittsburgh (MA)	...	...	...	12.7	15.9	20.0	...	...	...
Baltimore (SA)	...	...	...	...	...	...	28.8	13.4	18.0
Atlanta (SA)	...	...	...	...	...	...	...	25.4	32.7
Chicago (ENC)	...	...	...	...	...	...	...	...	14.7
Lesser centers	25.0	40.3	72.0	18.3	25.0	40.0	42.4	82.1	83.6

	EAST-SOUTH-CENTRAL			EAST-NORTH-CENTRAL			WEST-NORTH-CENTRAL		
	1900	1940	1965	1900	1940	1965	1900	1940	1965
New York (MA)	85.5	42.1	34.8	75.2	52.6	49.4	68.4	24.0	30.8
Atlanta (SA)	...	10.5	...	...	...	...	...	...	...
Cincinnati (ENC)	19.4	17.5	...	16.5	...	...	...	...	...
Chicago (ENC)	...	...	...	62.8	62.9	62.4	48.7	40.0	45.1
Cleveland (ENC)	...	...	...	10.7	11.2	...	...	...	...
Detroit (ENC)	...	...	...	...	...	11.8	...	...	...
Milwaukee (ENC)	...	...	...	...	24.1	21.2	...	...	...
Minneapolis (WNC)	...	...	...	...	...	...	17.7	27.2	36.3
New Orleans (WSC)	24.2	17.5	...	...	...	...	...	...	...
St. Louis (WSC)	21.0	...	...	...	...	...	13.9	13.6	14.3
Kansas City (WSC)	...	...	...	...	...	...	15.2	24.8	30.8
Lesser centers	64.5	93.0	93.9	27.3	47.4	55.3	55.1	71.2	78.0

	WEST-SOUTH-CENTRAL			MOUNTAIN			PACIFIC		
	1900	1940	1965	1900	1940	1965	1900	1940	1965
New York (MA)	86.4	36.2	36.3	91.7	59.7	22.4	94.8	33.9	10.3
Chicago (ENC)	11.9	...	10.9	51.7	35.5	10.3	34.5	13.6	...
Minneapolis (WNC)	...	...	...	...	14.5	...	...	...	...
New Orleans (WSC)	27.1	...	12.7	...	...	...	...	...	...
St. Louis (WSC)	61.0	20.7	12.7	...	...	...	...	...	...
Kansas City (WSC)	25.4	31.0	14.5	...	27.4	...	...	...	...
Houston (WSC)	...	24.1	29.0	...	...	...	...	...	...
Dallas (WSC)	13.6	27.6	54.5	...	...	...	...	...	...
San Francisco (P)	...	...	...	23.3	19.4	...	48.3	64.4	51.5
Los Angeles (P)	...	...	...	...	17.7	...	...	25.4	27.9
Seattle (P)	...	...	...	...	...	...	...	11.9	...
Lesser centers	23.7	86.2	80.0	63.3	90.3	96.6	25.9	30.5	26.5

Note: Major centers, identified in Table 12-1, are shown here only if they served as correspondents for 10 percent of the banks at some date; "lesser centers" includes all places except major centers.

spondent center for the south Atlantic area. Lesser centers, some of which are located within the region, have become increasingly frequent choices as correspondent-bank sites by the banks of the south Atlantic states.

The east-south-central states provide evidence of an even more drastic decline in the positions of cities that once were of considerable significance to the region. St. Louis has all but disappeared as a choice for bank correspondents, and the attractiveness of New Orleans and Cincinnati has decreased sharply in recent decades. These old centers which once served the region were not located within it, but were found near its edge. Here, as in the south Atlantic states, lesser centers have become increasingly frequent choices as correspondent-bank sites; and again it seems likely that many growing financial centers lie within the region itself.

In both the east- and west-north-central states, New York's correspondent role has declined; but Chicago's position has remained remarkably stable. By the time of World War II, Milwaukee had become a relatively important correspondent center for banks in the east-north-central states and, in particular, for banks in the state of Wisconsin. Minneapolis, a frequent choice of banks in the Dakotas as well as Minnesota, and Kansas City have steadily gained ground in the west-north-central states.

The changing regional orientation that a center may develop over the years is illustrated by Kansas City's role as a correspondent center for mid-continent banks. Its growing strength in the west-north-central states both before and after World War II received comment above. In the west-south-central and mountain states, Kansas City gained ground between 1900 and 1940, but failed to maintain its position after World War II. In fact, it is in this section of the nation that some of the sharpest shifts in position, both upward and downward, are to be found. New York had had unambiguous primacy in the early 1900s, but its position had declined precipitously by mid-century. St. Louis, a very active early correspondent center for banks in the west-south-central states, has fallen to a minor position. Chicago's position deteriorated in the mountain states.

The decline in dependence on outlying centers for correspondent functions in mid-continent has been accompanied by a spectacular growth of banking institutions within the region. Dallas and

Houston have expanded their services to banks in the west-south-central states significantly. New Orleans has not shared in the growth of financial independence within the region, however, indicating that the growth of regional financial self-sufficiency need not work to the advantage of established centers in the region. Nearly all banks in the mountain states select correspondents in cities other than the major centers identified here, including such important regional centers as Denver and Salt Lake City. Moreover, chain banking is widespread in the area. Nearly all banks in Arizona, for example, are branches of institutions headquartered in either Phoenix or Tucson. Boise and other Idaho communities have extensive chains in their state, and branches of banks headquartered in Salt Lake City and Ogden can be found in all parts of Utah.

In the area along the Pacific coast, New York's decline has been nothing short of spectacular. San Francisco, a regional center, has attained first rank by virtue of New York's waning importance, rather than by any substantial increase in its own correspondent activities. Growth in another regional center, Los Angeles, has shifted it upward in rank above New York and another outlying center of past importance, Chicago. Lesser centers are seldom selected by Pacific banks as correspondent-bank sites. The chain banking, now prevalent in the state of California, influences the selection pattern; for several major California banks are known to maintain headquarters in both San Francisco and Los Angeles.

CORRESPONDENT RELATIONS AMONG THE MAJOR CENTERS

The correspondent role of New York has not declined as greatly for banks in the major cities as for the nation as a whole. Nearly 85 percent of the banks in the other major centers still maintain at least one correspondent in New York, as compared with 95 percent at the turn of the century. To be sure, a number of other centers have gained in importance, but not necessarily at the expense of New York. The increasing number of correspondents used by banks has allowed old ties to be maintained while new ones were being established.

New York remains the most frequent selection as a correspondent-bank location in all but three centers; and in these places, Cincinnati, Detroit, and Milwaukee, it is tied with Chicago for first place. In 1900, however, the proportions of institutions with New York correspondents had ranged upward from a low of 84 percent in Minneapolis to a high of 100 percent in thirteen centers. By 1965 the range was from 43 percent in Boston to 100 percent in six centers, all relatively old and in the eastern half of the nation—New Orleans, Cincinnati, St. Louis, Cleveland, Buffalo, and Washington. (The low percentage in Boston arises from the inclusion of numerous savings banks located there.)

Although overall a denser network of intercenter correspondent ties had developed, some older financial centers have declined in importance even with respect to big-city banks. Chicago's banks are still securely in second place as a correspondent for banks in other big cities, but the frequency with which they are chosen has fallen from three-fifths to two-fifths. Philadelphia and Boston each had served as correspondents for about a third of the banks in the other major centers in 1900, but they now are correspondents for only a seventh of the banks in the same places.

The frequency with which St. Louis is selected as a correspondent location is recorded as a seventh in both 1900 and 1965. On the other hand, San Francisco, Los Angeles, and Dallas which had been chosen infrequently at the turn of the century are currently selected as a correspondent location by a sixth of the banks in the other major centers. An increase in frequency of selection is recorded for each remaining center although none is chosen by as many as a seventh of the "big-city" banks (see Table 12-3).

The pattern of intercenter correspondent relations in 1900 approximates the notion of a hierarchy. Following a suggestion by Winsborough and others (1966) with respect to quantification of the concept of a hierarchy, we first examined the frequency with which each center was selected as a correspondent-bank location by each other center. An arbitrary criterion was introduced to identify the more important inter-center ties; specifically, each instance in which at least 25 percent of the banks in one center selected another given center as a correspondent-bank location is designated a "linkage" directed from the first to the second center.

As a first step in identifying centers at the top of the hierarchy, consider the number of linkages directed to each center. New York

Table 12-3. Percentage of Banks in Major Centers Selecting as a Principal Correspondent a Bank Located in a Major Center in 1900 and 1965

Location of Correspondent	1900	1965	Location of Correspondent	1900	1965
Atlantic Seaboard			Transmontane East—cont.		
Boston	32.3	14.6	Detroit	0.2	9.2
New York	95.3	83.2	Milwaukee	0.6	4.3
Philadelphia	34.2	14.7			
Baltimore	4.5	7.1	Mid-Continent		
Washington	...	6.4	St. Louis	13.1	14.4
Miami	...	3.5	Minneapolis	1.0	7.4
			Kansas City	2.2	11.5
Transmontane East			Houston	0.2	5.2
New Orleans	0.6	5.4	Dallas	...	17.0
Cincinnati	2.5	3.1			
Chicago	63.3	42.3	Far West		
Pittsburgh	0.7	10.1	San Francisco	5.7	16.2
Buffalo	...	3.5	Los Angeles	0.2	16.1
Cleveland	1.0	8.8	Seattle	...	4.1

is selected by each other major center; and Chicago is selected by each other major center except Miami. Boston is selected by ten other centers, including New York and Chicago. Philadelphia is selected by seven other centers, again including New York and Chicago. Boston is selected by Philadelphia, and 24 percent of Boston's banks choose Philadelphia. No other centers are selected so frequently, and no other centers are selected by New York, Chicago, Boston, or Philadelphia. The top of the hierarchy seems, then, to consist of New York, Chicago, Boston, and Philadelphia, with the former two centers closer to the apex.

A graphic representation of the important intercenter linkages other than those directed to New York or Chicago facilitates discussion (Figure 12-1). New York and Chicago are not shown in the representation, and the reader must visualize the directed arrows leading from virtually all major centers to these cities along with the directed arrows leading from them to Boston and Philadelphia. Second, Cleveland, Milwaukee, and Miami are not shown since no center had a linkage directed to them and their only directed linkages terminated in New York or Chicago. With respect to the

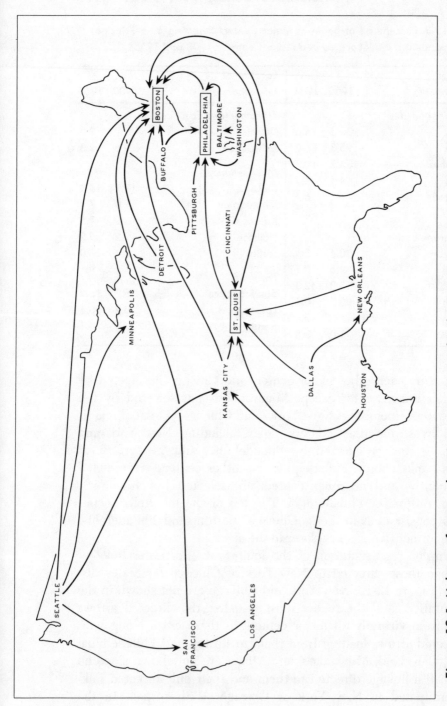

Figure 12-1. Linkages from each major center to each other major center which performed important correspondent-banking services for it in 1900, except New York and Chicago (see text for full explanation).

linkages represented, observe that only one place, St. Louis, had directed linkages to both Boston and Philadelphia, centers identified with the top level of the hierarchy. The other linkages directed to Boston originated in centers scattered over the nation; only proximate centers selected Philadelphia. St. Louis itself resembled Philadelphia in that it was selected only by neighboring centers. Kansas City and San Francisco appear as termination points in westerly regional networks somewhat less extensive than those centered on Philadelphia or St. Louis.

A much more complex set of interdependencies exist in 1965. To be sure, the very top of the hierarchy is orderly, perhaps even more so than in 1900. Fewer than 25 percent of New York's banks selected a correspondent in any other major center, but New York was selected as a correspondent location by more than 25 percent of the banks in each other major center. Chicago's only choice was New York, but it, in turn, was selected by sixteen major centers. Two other centers, Boston and Dallas, chose only New York although they were chosen by seven and eight other centers, respectively. The intercenter relations become much less clearly structured when the remaining linkages are examined, however.

There had been in 1900 relatively few reciprocal choices between centers. The five recorded involved Philadelphia, Boston, Chicago, and New York, centers toward the apex of the hierarchy. Fifteen reciprocal choices were found in 1965. None involved New York, Chicago, Boston, or Dallas. Philadelphia was involved in several reciprocal choices, however, suggesting that the positions of some fairly important banking centers have become more ambiguous.

There is no hope for representing graphically all the intercenter linkages observed in 1965, but the linkages terminating in Philadelphia, Boston, or Dallas can be displayed (Figure 12-2). The linkages for these three centers seem to represent the outcome of three different processes. Philadelphia's position became ambiguous; Boston maintained a rather strong position; and Dallas achieved national prominence.

At the turn of the century, New York and Chicago had selected Philadelphia as a correspondent-bank location; neither did so in 1965. Baltimore, Pittsburgh, and Washington selected Philadelphia in both years, but the choice was reciprocated by Philadelphia only

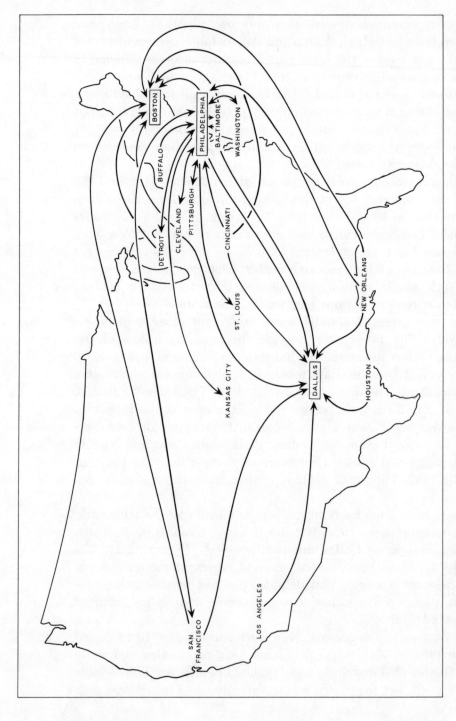

Figure 12-2. Linkages from each major center to Boston, Philadelphia, and Dallas for correspondent-banking services in 1965 (see text for full explanation).

in 1965. St. Louis had selected Philadelphia in 1900, but in 1965 Philadelphia selected St. Louis. A reciprocal linkage was recorded in 1965 between Philadelphia and Cleveland, Detroit, and San Francisco, respectively; at the earlier date, no tie had existed between Philadelphia and any of the three centers. At both dates, Buffalo selected Philadelphia; and Philadelphia, in turn, selected Boston. Although Philadelphia was selected by more centers in 1965 than at the turn of the century, it has become less clearly placed in the hierarchy.

Neither New York nor Chicago selected Boston in 1965 although both had done so in 1900; Boston itself no longer selected Chicago although it maintained a linkage directed to New York. All other ties between Boston and another major center were unreciprocated by Boston at both dates. Half the centers which had selected Boston at the turn of the century failed to do so in 1965, specifically, St. Louis, Buffalo, Kansas City, and Seattle. The other four centers, Philadelphia, Cincinnati, Detroit, and San Francisco, continued to choose Boston. Baltimore, New Orleans, and Cleveland which in 1900 had not chosen Boston did so in 1965, however. Overall Boston retains a strong correspondent role, serving major centers scattered through the nation.

Dallas, which was a choice of no other major center in 1900, was selected by eight centers in 1965; and Dallas no longer chose New Orleans, St. Louis, Chicago, or Kansas City as an important correspondent-bank location, but retained only its tie to New York. In fact, part of the expansion of correspondent services in Dallas occurred at the expense of St. Louis, to which it had been subordinate at the turn of the century. Houston no longer chose St. Louis although it had a linkage directed to Dallas. Although New Orleans and Cincinnati retained their tie to St. Louis, they also established a tie to Dallas. Dallas also carved out an important role among centers located in distant sections of the nation. By 1965 it performed important correspondent-bank functions for San Francisco and Los Angeles on the Pacific and for the eastern centers of Philadelphia, Baltimore, and Cleveland.

A skeleton financial network spanning the nation and focused on New York both directly and by way of Chicago had been in place by 1900; its outlines remained visible six decades later. Complementing this skeleton network in 1900 were denser regional networks within the long-settled parts of the nation, networks focused on such centers as Boston or Philadelphia, New Orleans, Cincinnati, or St. Louis. Banks in the more recently opened territory relied for services on the older extraregional centers of finance. By 1965 regional networks focused on newer centers had developed in the southern and western sections of the nation where population growth and industrial expansion were proceeding most rapidly. The rise of the new centers of finance presented no challenge to the older centers within the latter's proximate territory, but such extraregional significance as the older centers may have had came to an end. We appear to have witnessed the final stages of the continental colonialism which characterized the development of a national economy in the American setting.

Chapter 13

INDUSTRIAL STRUCTURE & A SYSTEM OF CENTERS

Perhaps the most ambitious study of a so-called metropolitan region undertaken to date resulted in a series of publications the last of which was entitled *Metropolis 1985* (Vernon, 1963). Since New York and its environs were the area under study, there can be no quarrel with the title of the concluding volume. By virtually any criterion, New York is the nation's most important city, the national metropolis.

In the classical formulation, commercial strength is taken as the hallmark of metropolis; and when great centers first appeared in Western Europe, they were distinguished by the breadth of their trade ties. Commerce then was one of the few lines of activity unambiguously identified with urban life, but in twentieth-century America, manufactures are as much a part of the urban scene as trade. By mid-century, special services oriented to near-national markets were giving impetus to urban growth. In such a national context, the foremost city is almost certain to evidence strength not only in trade, but in the manufacturing and service sectors as well. New York does so.

To adduce reasons for identifying San Francisco as the metropolis of the far west or Boston as the metropolis of New England and for withholding the designation metropolis from Los Angeles or

217

Detroit is likely to be a sterile exercise. Perhaps a more fruitful approach is to recognize the present potential for diversity in industrial structure among the nation's largest urban centers and to attempt to measure such variation as obtains. If there exists a system of major centers, differentiated and articulated nodes in the spatial structure of the national economy, it should be manifest in the diversity of industrial structures to be observed among the nation's major centers.

A convenient measure of intercenter differences in industrial structure is the index of dissimilarity, and the matrix of indexes calculated from distributions of the civilian employed over 148 detailed industry categories in 1960 is reproduced here for pairs of major centers (Table 13-1). A shift in the industrial attachment of no more than a fifth of the work force in Boston or Philadelphia would produce a match with the New York industrial structure. On the other hand, a redistribution by industry of at least three-tenths the work force in Pittburgh or Detroit would be required to reproduce the New York industrial structure. A scanning of the full matrix makes clear the fact that a description of diversity in terms solely of the degree to which each center's structure departs from the New York structure fails to abstract all pertinent information in the matrix, however. Philadelphia, for example, is no more dissimilar to Cincinnati, St. Louis, and Chicago than to New York in industrial structure; but Boston is unambiguously more dissimilar to these three centers of the interior than to New York. Pittsburgh and Detroit differ from one another with respect to industrial structure as sharply as each differs from New York.

Grouping centers by "when and where" they grew up helps to structure the overall pattern of intercenter differences, but the size of matrix and complexity of pattern make it difficult to grasp the overall configuration. A graphic representation of the matrix is offered as an aid (Figure 13-1). Note, for example, that the configuration of points representing New York, Pittsburgh, and Detroit is triangular; each differs sharply from the others with respect to industrial structure. Philadelphia lies much closer to the Cincinnati-St. Louis-Chicago cluster than does Boston, whose position in the two-space remains close to New York. (The graphic representation is based on results of the Guttman-Lingoes "smallest space" analysis of the matrix appearing in Table 13-1. In the configuration dis-

Table 13-1. Indexes of Dissimilarity with Respect to 1960 Detailed Industrial Structure Between Specified Pairs of Major Centers

Center	Bo	Pa	Ba	Ci	SL	Cg	Pi	Bu	Cl	Dt	MI	NO	SF	Mp	KC	LA	Ho	Dl	Se	Mi
NY	17	18	22	21	22	22	31	27	24	33	28	27	20	23	23	20	29	23	29	28
Bo	—	18	22	23	22	20	28	26	24	33	23	26	21	22	24	20	30	23	30	29
Pa	:	—	17	18	16	16	24	21	19	27	22	25	20	22	19	18	25	23	29	30
Ba	:	:	—	19	18	21	22	17	21	30	28	23	19	23	22	21	26	21	26	29
Ci	:	:	:	—	13	17	27	18	17	26	22	27	23	19	17	20	23	22	27	30
SL	:	:	:	:	—	17	23	18	18	27	23	27	22	20	15	19	25	20	27	32
Cg	:	:	:	:	:	—	22	21	16	27	21	29	24	20	18	21	27	24	32	33
Pi	:	:	:	:	:	:	—	20	23	31	26	32	29	27	27	28	30	30	34	36
Bu	:	:	:	:	:	:	:	—	16	25	23	32	26	24	24	24	27	27	31	34
Cl	:	:	:	:	:	:	:	:	—	20	18	31	26	24	22	23	28	26	32	35
Dt	:	:	:	:	:	:	:	:	:	—	26	35	30	29	28	29	30	33	34	36
MI	:	:	:	:	:	:	:	:	:	:	—	34	30	24	27	27	32	30	34	35
NO	:	:	:	:	:	:	:	:	:	:	:	—	20	25	24	28	20	23	26	23
SF	:	:	:	:	:	:	:	:	:	:	:	:	—	19	18	18	23	21	19	24
Mp	:	:	:	:	:	:	:	:	:	:	:	:	:	—	15	22	24	21	25	29
KC	:	:	:	:	:	:	:	:	:	:	:	:	:	:	—	22	23	19	25	26
LA	:	:	:	:	:	:	:	:	:	:	:	:	:	:	:	—	26	18	20	24
Ho	:	:	:	:	:	:	:	:	:	:	:	:	:	:	:	:	—	20	26	25
Dl	:	:	:	:	:	:	:	:	:	:	:	:	:	:	:	:	:	—	24	21
Se	:	:	:	:	:	:	:	:	:	:	:	:	:	:	:	:	:	:	—	27

Note: Detailed industrial structure measured by distribution of the civilian employed over 148 industry categories. Full identification of caption and stub symbols appears in Tables 10-3 and 10-9.

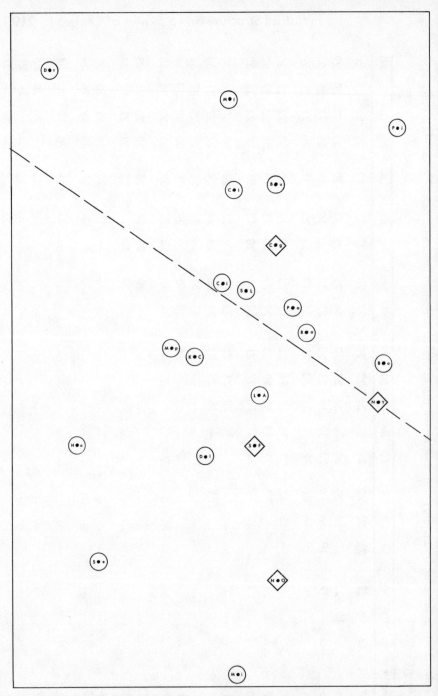

Figure 13-1. Dissimilarity in industrial structure among major centers, except Washington, in 1960.

played in Figure 13-1, the "distance" of each center from each other center approximates the corresponding "distance" in the matrix, measured as rank order with respect to the index of dissimilarity, as closely as possible in a two-space.)

We prefer not to label the two dimensions along which the centers are arrayed; neither do we wish to delimit clusters of centers and designate them as to type. We have arbitrarily oriented the configuration with respect to New York, New Orleans, Chicago, and San Francisco, the prime metropolises of the east, the south center, the north center, and the west as these sections of the nation were settled and brought into exploitation. For expository convenience, we have sketched on the configuration a "diagonal" passing through the position in the two-space occupied by New York.

Just above the so-called diagonal lie Boston, Baltimore, Philadelphia, St. Louis, and Cincinnati; above them lies Chicago; still further above the diagonal lie Pittsburgh, Buffalo, Cleveland, Milwaukee, and Detroit. These are centers of the type referred to by Duncan et al. (1960: 271) as "diversified-manufacturing" centers. The group falling closest to the diagonal is made up of centers which, we have argued, were regional metropolises before the industrialization of manufactures and, then, became also the sites of large-scale manufacturing concentrations. The group further from the diagonal includes the centers which gained national prominence with the industrialization of manufactures.

Below the so-called diagonal cluster the centers identified by Duncan and his colleagues (1960: 271) as "regional metropolises and capitals." More remote from the diagonal are the newest of the regional centers; still more remote is the "special case" of Miami.

Verbal descriptions purporting to "explain" the configuration tend quickly to enter the realm of fantasy. Each viewer will perceive positioning of centers in the two-space which is significant to him. One may be struck, for example, by the alignment of Pittsburgh, Boston, New York, and New Orleans and choose to argue that these are the prototypes of a center of large-scale manufacture, a regional metropolis wherein industrial manufactures located, a national metropolis, and a regional metropolis wherein commerce remains preeminent in the industrial mix. Another viewer may be struck by the similarity of Baltimore to Philadelphia, of St. Louis to Cincinnati, of Kansas City to Minneapolis and relate it to a corresponding similarity in their locations within the nation, the timing of their growth, or their positions on the transport network. Still another viewer may note that the "distances" separating, say, Boston, Baltimore-Philadelphia, and St. Louis-Cincinnati tend to parallel their respective differences in "age." Such hunches can, at best, only generate leads for further exploration.

The evidence we have on the development of a system of differentiated centers within the United States has been presented fully, and we can offer no sounder forecast than our reader can. To foreclose the possibility that the new centers will retain their distinctive character seems unwarranted to us, however. Should distinctiveness persist, we suspect that some years hence the line of argument may be that air transport so drastically changed the scale of distance that a center could once again gain national prominence without incorporating into its industrial structure the complement of activities associated with the manufacturing center.

PART III

SPECIAL STUDIES

Chapter 14

THE MARKET FOR MANUFACTURED GOODS

Materials have become available through the Commodity Transportation Survey of the 1963 Census of Transportation, conducted by the Bureau of the Census, which permit some further exploration of the center-region relations relevant to the distribution of manufactured goods from the production establishment. The classic formulation of the metropolis-region relation posits a set of major urban centers each with a continuous hinterland that is organized exclusively by units at the urban core and includes important supply and market areas for the center's industrial specialties. By searching the monographic and periodical literature, Duncan et al. (1960: ch. 11) attempted to ascertain the origin of inputs and and destination of outputs for each major center's manufacturing specialties. They found the proportion of metropolitan industrial specialties for which the classic conception of the center-region relation was relevant to be astonishingly low. We have not attempted to replicate their work, but we have examined the data of the Commodity Transportation Survey which relate to the destinations of goods manufactured in the major centers.

A sample of manufacturing plants in the nation was selected for the survey, and a sample of their shipping records provides the basic data. Slightly more than a million shipping records were drawn from the files of about 10,000 manufacturing establishments in the conterminous United States. All types of commodities shipped intercity by manufacturers and all modes of transportation except pipeline were covered in the survey

That part of the Commodity Transportation Survey bearing most directly on the center-region relation is the "Production-Area" report (1963 Census of Transportation, Vol. 3). Twenty-five "production areas," each including one or more metropolitan centers, were delineated. Four production areas consist of groups of metropolitan areas each smaller in population size than the centers on which we have focused attention; we have not examined the data for these areas individually. The New York State and New Jersey parts of the metropolitan complex centered on New York City are treated as distinct production areas. Each other production area includes one of our so-called major centers, often in combination with a lesser metropolitan area; and these we will refer to by the name of the major center. The national territory is exhausted by the twenty-five production areas, "other" metropolitan locations, and nonmetropolitan locations.

The units in which intercity shipments have been measured are tons, and distributions of tonnage originating in each production area are reported by transport mode, by length of haul, and by destination according to the twenty-sevenfold areal classification just outlined. When commodities measured in tons are aggregated, the total is not necessarily heavily weighted with the commodities of high value. Neither are the commodities that weigh heavily in the total necessarily outputs of industries identified as local manufacturing specialties.

Two features of the data limit their usefulness for the study of center-region relations. First, intracity shipments have been excluded from the survey data. Hence, short-haul shipments are underestimated; and the degree of underestimation must vary among areas depending upon the delimitation of the metropolitan complex, the political fragmentation within the complex, and the unevenness with which manufacturers are distributed through the complex. Second, the port of exit was defined as the destination for shipments to foreign markets. Hence, for some port centers, the importance of the local market may be overstated. Anomalies in our findings often seem to reflect these features of the basic data set.

Airborne shipments were insignificant in the distribution of commodities (tonnage basis) from each major center for the aggregate of manufactured goods although air transport was rather common for particular commodities. Waterborne movements weighed heavily in shipments originating in the New Jersey part of the New York complex, the Philadelphia area, San Francisco, and Houston. Coastal shipping of petroleum products from these centers, each of which had refineries in the vicinity, appears to account for most of the water-borne movements. Waterborne shipments from other seaports appear relatively unimportant, perhaps because the tonnage destined for foreign markets is high relative to the tonnage destined for the domestic market.

In most areas, manufacturers rely on a combination of rail and motor transportation to move their products intercity. Rail shipments, including "piggy back," were especially important in the distribution of manufactured goods from Pittsburgh, Buffalo, Cleveland, Milwaukee, and the "Motor City" of Detroit. These are the places whose rise to national prominence we earlier identified with the steam-and-steel complex. (See table 14-1)

Although we know the proportion of intercity shipments terminating within a hundred miles of the point of origin, we cannot estimate satisfactorily the importance of the local "metropolitan-area" market for manufacturers. The territory designated the metropolitan area by McKenzie (1933) would fall within a hundred miles of the center, for daily contact between the population residing in the territory and the core of the metropolitan community is a necessary characteristic (Duncan et al., 1960: 90 ff.). Short-haul intercity movements are most important in the distribution of goods from New York, Philadelphia, San Francisco, Los Angeles, and Seattle. Each is a port center, however; and we may only be detecting the effect of the port-of-exit rule.

The territorial extent of the "metropolitan region" associated with a center is more difficult to specify in terms of distance from the metropolitan core than is the extent of the metropolitan area. A center's region generally is conceived as the territory within which the resident population is oriented to the center as a market-

Table 14.1. Manufacturers' Intercity Shipments (tonnage) by Transport Mode, by Miles Shipped, and by Destination, for Selected Large Centers, 1963

Center (Production Area)	TRANSPORT MODE			MILES SHIPPED			Metropolitan	DESTINATION Centers Receiving 4 percent
	Water	Rail	Motor	Under 100	100–499	Over 499		
	Percent of Tons							
Boston	8	12	79	42	38	20	71	NY–NY;NY–NJ
New York, N.Y.	1	9	86	56	28	16	84	Bo;NY–NJ;Pa
New York, N.J.	26	15	58	52	42	6	81	NY–NY;Pa
Philadelphia	58	12	30	69	26	5	90	Bo;NY–NJ
Baltimore	13	21	66	36	49	15	72	NY–NY;NY–NJ;Pa
Cincinnati	6	30	64	28	51	21	75	NY–NY;Cg;Dt
St. Louis	23	32	44	28	56	16	66	Cg
Chicago	10	35	54	39	45	16	80	Dt;Ml
Pittsburgh	10	52	38	27	59	14	78	Cg;Cl;Dt
Buffalo	8	45	45	17	71	12	77	NY–NJ;Cg;Cl;Dt
Cleveland	1	43	55	27	58	15	74	Cg;Pi;Dt
Detroit	3	46	51	38	45	17	79	Cl
Milwaukee	0	54	44	17	41	42	64	Cg;Dt
San Francisco	30	28	41	54	22	24	75	LA
Minneapolis	8	36	50	38	48	14	49	Cg
Los Angeles	19	16	64	57	29	14	86	NY–NJ;SF
Houston	92	4	4	4	9	87	89	Bo;NY–NY;NY–NJ;Pa;Ba
Dallas	0	26	74	24	57	19	70	Ho
Seattle	13	26	61	56	15	29	61	SF;LA
Atlanta	0	23	75	21	59	20	55	SF;LA
Denver	0	33	66	23	37	40	56	Cg;LA

Note: Production area includes metropolitan area of center listed and may include one or more metropolitan areas of lesser centers. Component parts of each production area are identified in 1963 Census of Transportation, place for commodities in interregional exchange and as a supplier of capital, investment opportunities, and specialized services. Delimitations of metropolitan regions proposed by investigators typically depict rather compact regions in the northeastern part of the nation where large urban centers are closely spaced and more areally extensive regions in other sections of the nation.

We assumed that shipments originating in metropolitan centers which were destined for customers or primary distributors located within a hundred miles of the point of origin were "local." By and large, goods destined for a location at least 500 miles away can be assumed to move beyond the limits of the metropolitan region in which production occurred. How important is the "metropolitan-region" market, the market in territory 100 to 500 miles distant? The distance-shipped data alone suggest that it is more important than the local or the extraregional market for manufacturers in a number of centers. Included among the centers are, however, such places as Pittsburgh, Buffalo, Cleveland, and Detroit, centers in which the distinctively metropolitan function appeared rather poorly developed. Goods moving from these manufacturing-belt centers, as well as from Baltimore, Cincinnati, and Chicago, are typically destined for other metropolitan locations, not for locations in the nonmetropolitan hinterland associated with the notion of the metropolitan region. Taking into account both distance shipped and metropolitan status of the destination, a slightly stronger case for a regional orientation can be made with respect to manufacturers in St. Louis and Dallas, a still stronger case with respect to manufacturers in Atlanta and Minneapolis. Overall, however, it is the intermetropolitan flows that are most salient in the distribution pattern.

New York and Chicago appear most frequently in the list of metropolitan markets (separately identified production areas) into which manufacturers in metropolitan complexes ship a substantial volume of goods. There is relatively little overlap in the list of centers oriented to New York and Chicago, respectively, as a marketplace, however. New York is an important destination for flows originating in Boston, Philadelphia and Baltimore, neighboring ports on the Atlantic, in Houston on the Gulf, and in Los Angeles on the Pacific. Chicago serves as a major market for manufacturers in the trans-Appalachian centers of Pittsburgh, Cleveland, and Milwaukee and the trans-Mississippi centers of St. Louis, Minneapolis, and Denver. Only manufacturers in Cincinnati and Buffalo direct major flows of goods into both the New York and Chicago markets.

Equally distinctive sets of metropolitan producers are oriented to the Philadelphia, Detroit, and Los Angeles markets. Manufacturers in neighboring New York and Baltimore and in the Gulf port of Houston direct a sizable share of their intercity shipments to the Philadelphia market. The Detroit market is a focus for flows originating in the manufacturing-belt centers of Cincinnati, Chicago, Pittsburgh, Buffalo, Cleveland, and Milwaukee. Los Angeles is the destination of substantial flows originating in the more westerly cities of San Francisco, Seattle, and Denver.

The destinations of the outputs of local manufacturing specialties can be ascertained from the data collected through the Commodity Transportation Survey for several metropolitan complexes. There is some slippage in matching the survey data with a metropolitan industrial specialty as identified in Chapter 10, but we can effect a number of more or less reasonable matches. (The delimitation of the metropolitan area typically differs, and the industry category may differ in level of aggregation from the commodity category.) For each "matched" manufacturing specialty, we have compiled some summary statistics on the mode of transportation, the distance shipped, and the relative importance of metropolitan markets.

The distribution of outputs of one Boston manufacturing specialty with a close tie to resource extraction, the confectionary industry, can be determined. The domestic market appears to approximate the national territory insofar as its spatial features are revealed by the summary statistics (Table 14-2). Although two-thirds of the tonnage leaves the plant by motor carrier or private truck, typically short-haul carriers, most of the output enters long-distance flows. The immediately surrounding territory provides a market for less than a tenth of the output. Three-fifths is channeled into other Production Areas, among which New York, Chicago, Cleveland, San Francisco, and Los Angeles are important receiving areas. About a fifth of the intercity shipments are directed to metropolitan areas which lie outside the Production Areas, and the mean distance of this other-metropolitan flow is in excess of 1,000 miles. A seventh of the output moves into the nonmetropolitan market area, and the mean distance associated with the nonmetropolitan flow is nearly 600 miles.

In contrast, the market area for paper manufacturers in the Boston complex appears spatially constricted. Although the flows are of relatively short length, they are primarily inter-metropolitan in character. For example, only a tenth of the output is shipped as far as 500 miles from the plant; and the mean distance of flows directed to the other-metropolitan and non-metropolitan markets is less than 300 miles. On the other hand, four-fifths of the shipments are destined for a metropolitan location: a fifth within the Boston complex; a third to the New York complex; a sixth to production areas other than New York; and a tenth to metropolitan territory outside the production areas.

The flows generated by Boston manufacturers in four other specialty lines, which like the paper industry have only indirect ties with resource extraction, also fail to evoke the imagery of a market area coextensive with the surrounding territory organized by the center.

A scanning of the destinations of outputs of manufacturing specialties in New York and Philadelphia reveals instances of a near national market area and of a constricted market area in which intermetropolitan flows are prominent. The Philadelphia textile industry perhaps offers the closest approximation to a center-generated flow destined for proximate nonmetropolitan territory. About two-thirds of the textile output is directed to nonmetropolitan locations with a mean distance of something under 400 miles from the point of origin. Overall, the case for a regional orientation on the part of manufacturers in the metropolitan complexes bordering the Atlantic is weak, however.

Although comparisons among centers with respect to the distributional pattern are fraught with ambiguity, proximate nonmetropolitan territory appears to be a more important market area for manufacturers in the specialty industries of some interior centers than for their counterparts in the metropolitan complexes around the Atlantic ports. Each St. Louis specialty for which destinations can be ascertained directs a substantial share of its products into proximate nonmetropolitan territory (Table 14-3). The nonmetropolitan flow accounts for a quarter to a half of the intercity ship-

Table 14-2. Characteristics of Manufacturers' Intercity Shipments (tonnage) from Boston, New York, and Philadelphia Metropolitan Complexes, 1963

Complex and Commodity by Resource-Use Stage	SIC Code	TRANSPORT MODE				MILES SHIPPED		Production Area				DESTINATION		Centers Receiving 4 percent
		Wa-ter	Rail	Mo-tor	Air	Under 100	Over 499	Same	Other	Other Met- ro.	Non-Met- ro.	Other Met- ro.	Non-Met- ro.	
										Percentage		Miles		
BOSTON														
First stage														
Confectionary	207	0	33	66	0	7	68	3	60	23	14	1,090	570	NY–NY;Cg:Cl;SF;LA
Other														
Paper and allied products	26	0	7	93	0	37	10	19	52	10	18	290	270	NY–NY;NY–NJ
Rubber and plastic	30	0	11	83	0	38	36	23	47	18	13	1,050	620	NY–NY;Pa
Leather products	31	0	22	76	0	19	38	8	54	18	20	510	380	NY–NY;SL
Electrical machinery	36	0	7	80	1	37	27	17	50	18	15	620	390	NY–NY
Instruments	38	2	3	72	2	8	46	6	64	16	14	1,130	670	NY–NY;NY–NJ;LA
NEW YORK														
Other														
Apparel (NY)	23	0	4	65	3	20	49	4	61	18	16	880	680	Bo;NY–NJ;Pa;Cg;LA

Apparel (NJ)	23	15	38	20	6	38	36	6	60	24	10	910	830	Bo;NY–NY;Pa
Drugs (NY)	283	2	6	60	0	34	50	4	58	10	28	1,140	430	NY–NJ;LA;At
Drugs (NJ)	283	17	44	12	3	44	41	6	78	12	3	500	560	NY–NY;Pa;Cg;LA
Leather (NY)	31	0	24	58	1	29	45	13	56	19	11	910	480	Bo;NY–NJ;Cg;LA
Instruments (NY)	38	0	11	65	2	26	48	8	60	19	13	1,290	420	Bo;NY–NJ;Pa;Cg;LA
Instruments (NJ)	38	1	2	79	1	30	37	8	58	24	10	1,240	480	NY–NY;Pa;Cg;Pi;Bu
Miscellaneous (NY)	39	0	18	66	0	23	56	14	62	12	11	830	860	Cg;LA
Miscellaneous (NJ)	39	13	67	2	3	67	28	0	89	8	2	890	570	Pa;SF;At
PHILADELPHIA														
First stage														
Confectionary	207	5	2	93	0	34	32	2	75	14	9	460	320	Bo;NY–NY;Ba;Ci;Cg;Ml
Petroleum	291	83	3	13	0	81	1	39	54	2	5	170	140	Bo;NY–NJ
Other														
Textiles	22	0	1	96	0	20	18	3	27	3	67	740	370	NY–NY;NY–NJ
Apparel	23	1	0	54	1	33	32	2	56	13	29	940	500	NY–NY;NY–NJ;Cg
Paper and allied products	26	0	12	87	0	55	7	20	61	5	14	370	130	NY–NY;NY–NJ
Drugs	283	1	13	84	0	10	69	1	88	9	2	1,020	690	NY–NJ;Cg;Pi;SF;At
Fabricated metal products	34	1	40	59	0	42	18	13	49	11	26	680	290	Bo;NY–NY;NY–NJ
Electrical machinery	36	1	24	69	2	22	49	6	50	20	23	920	680	NY–NY;NY–NJ;LA
Instruments	38	1	0	83	3	28	31	8	46	26	20	520	440	NY–NJ;Dt;LA

Table 14-3. Characteristics of Manufacturers' Intercity Shipments (tonnage) from Cincinnati, St. Louis, and Chicago Metropolitan Complexes, 1963

Complex and Commodity by Resource-Use Stage	SIC Code	TRANSPORT MODE				MILES SHIPPED		Production Area		Other Met-ro.	Non-Met-ro.	DESTINATION		Centers Receiving 4 percent
		Wa-ter	Rail	Mo-tor	Air	Under 100	Over 499	Same	Other			Other Met-ro.	Non-Met-ro.	
						Percentage						Miles		
CINCINNATI														
First stage														
Beverage industries	208	0	35	64	0	38	20	26	22	23	29	360	230	
Other														
Chemicals	28	0	35	64	0	12	28	4	58	27	12	330	300	NY–NJ;Ba;Cg;Pi;Cl
Rubber and plastic	30	0	12	86	0	7	42	4	38	36	23	620	540	Pa;Bu;Mp;At
Nonelectrical machinery	35	0	21	77	0	36	35	2	30	19	49	640	200	NY–NY
Motor vehicles	371	0	46	53	0	6	40	2	62	23	13	600	440	NY–NY;NY–NJ;Dt
ST. LOUIS														
First stage														
Food and kindred products	20	0	46	54	0	12	34	4	27	30	38	440	370	Cg

Petroleum	291	63	9	28	0	34	1	23	8	38	31	200	140	Cg

Other

Chemicals	28	0	55	44	0	23	28	18	33	23	26	470	370	Ci;Cg
Clay and glass	32	0	42	58	0	48	8	25	11	18	46	240	140	Cg
R.R. equipment	374	0	93	7	0	8	30	0	27	42	31	500	390	Cg

CHICAGO

First stage

Confectionary	207	0	35	61	0	7	65	5	52	23	20	590	620	SF;LA
Abrasives	329	0	50	49	0	61	17	53	25	8	14	420	260	NY–NY;Dt

Other

Paints	285	0	7	93	0	17	40	11	29	38	22	530	360	SL;Dt;Ml
Paving materials	295	0	25	74	0	29	10	12	36	21	31	230	230	
Rubber and plastic	30	0	1	97	0	13	39	7	48	31	14	450	420	NY–NY;SL;Pi;Cl;Dt
Steel works	331	5	48	46	0	47	10	36	29	19	16	260	290	Dt;Ml
Fabricated metal products	34	4	52	42	0	24	30	16	36	29	18	410	400	Dt
Nonelectrical machinery	35	0	35	60	0	23	37	17	41	19	24	590	460	NY–NY;Dt
Electrical machinery	36	0	22	61	1	13	55	9	50	30	12	630	540	Bo

ments generated by these industries and has associated with it a mean distance in the range of 100 to 400 miles. Strong flows of this type also characterize manufacturers of beverages in Cincinnati, manufacturers of paving materials in Chicago, and manufacturers of nonelectrical machinery in both metropolitan complexes.

Not only do the specialties of St. Louis reveal a relatively strong market orientation to the surrounding nonmetropolitan territory; their orientation to major metropolitan markets also is distinctive in that it is focused almost exclusively on Chicago. In contrast, manufacturers in the specialty lines of the metropolitan complexes on the Atlantic and of Cincinnati and Chicago direct substantial flows to several major metropolitan markets, some neighboring centers and others located at considerable distances from the producing center. Since the coverage of each center's specialties is incomplete, we cannot claim to have captured all salient features of the market orientation of manufacturers in the lines of industry which are local specialties; but something of the diversity of market orientations is revealed.

The destinations of outputs of several manufacturing specialties in each of the five trans-Appalachian centers whose rise to national prominence coincided with the industrialization of manufactures can be ascertained. A significant feature of the distribution pattern is the frequency with which one center serves as a market for the products of another center although each is a center of large-scale manufactures (Table 14-4). Buffalo and Cleveland are major destinations of flows generated by Pittsburgh manufacturers of clay and glass products; products of steel works, iron and steel castings, nonferrous shapes, and structural metal are shipped from Pittsburgh plants to Cleveland; and from electrical-machinery manufacturers in the Pittsburgh area, a substantial volume of goods flows into the Buffalo area. Among the other flows between manufacturing centers recorded are: outputs of the rubber-and-plastic and primary-metal industries in Buffalo moving to Cleveland and Detroit; outputs of the Buffalo transportation-equipment industry en route to Detroit; shipments from manufacturers of paints, miscellaneous primary metal products, fabricated metal products, electrical machinery, and motor vehicles in Cleveland moving to Detroit; iron and steel castings and nonferrous shapes and castings moving from Cleveland to Buffalo and Detroit; outputs of steel works and the nonelectrical-machinery industry moving from Cleveland to Pittsburgh and Detroit; cutlery and tools, bolts and nuts, and nonelectrical machinery en route from Detroit manufacturers to the Cleveland area; and instruments produced by Milwaukee manufacturers in transit to Detroit.

Manufacturers in the specialty industries of these manufacturing centers also tap the major metropolitan markets associated with the encircling older centers. Manufacturers in Pittsburgh and Buffalo are suppliers to the market centered on Boston; those in Cleveland and Detroit are suppliers to the St. Louis market. Products manufactured in Pittsburgh, Buffalo, and Cleveland are shipped to Baltimore, and flows are generated from Buffalo, Detroit, and Milwaukee to Philadelphia. Substantial shares of the outputs of the key industries in each manufacturing center are destined for New York, Cincinnati, and Chicago.

More distant major metropolitan markets appear to be tapped more selectively by manufacturers in the manufacturing centers. Outputs of the electrical-machinery industry move from Pittsburgh to Dallas and Seattle, from Cleveland to San Francisco and Los Angeles, and from Milwaukee to Dallas. Directed to Atlanta are flows from Buffalo manufacturers of transportation equipment and from Milwaukee manufacturers of chemicals and instruments. Directed to Los Angeles are flows from Milwaukee manufacturers of chemicals and Detroit manufacturers of cutlery and tools and motor vehicles.

The market orientations of manufacturers in a key industry can be ascertained for only four centers located outside the northeast, but four distinctive distribution patterns are revealed (Table 14-5). Combined in San Francisco are a few specialties oriented to the market along the Pacific and one specialty with an unambiguously national market orientation. The major metropolitan markets tapped by the San Francisco canning industry span the continent: New York, Philadelphia, Chicago, and Los Angeles are major destinations for the four-fifths of output moving to distant metropolitan markets. The Minneapolis food industry, on the other hand, is oriented to a much more constricted market area. Half the outputs are directed into nonmetropolitan territory, and the mean distance associated with the flow is just under 200 miles. A more dispersed market area is tapped by the Minneapolis manufacturers of nonelectrical-ma-

Table 14-4. Characteristics of Manufacturers' Intercity Shipments (tonnage) from Pittsburgh, Buffalo, Cleveland, Detroit, and Milwaukee Metropolitan Complexes, 1963

Complex and Commodity by Resource-Use Stage	SIC Code	TRANSPORT MODE				MILES SHIPPED		Production Area		Other Met- ro.	Non- Met- ro.	DESTINATION		Centers Receiving 4 percent
		Wa- ter	Rail	Mo- tor	Air	Under 100	Over 499	Same	Other			Other Met- ro.	Non- Met- ro.	
		Percentage										Miles		
PITTSBURGH														
Other														
Clay and glass	32	0	21	77	0	11	5	2	57	11	30	350	200	Ba;Ci;Bu;Cl
Steel works	331	14	56	30	0	24	16	11	56	13	21	490	280	Ci;Cg;Cl;Dt
Iron and steel castings	332	0	34	66	0	44	3	10	60	11	19	260	210	NY–NJ;Cg;Cl
Nonferrous shapes	335	0	19	79	0	20	5	12	49	22	17	210	240	Ci;Cg;Cl
Structural metal	344	2	78	20	0	40	8	31	29	23	17	280	350	Cl
Electrical machinery	36	2	29	64	0	6	40	4	58	19	19	690	520	Bo;Ci;Cg;Bu;Dl;Se
BUFFALO														
First stage														
Food and kindred products	20	0	39	60	0	24	8	10	49	14	27	540	170	Bo;NY–NY;NY–NJ; Pa;Cl
Other														
Chemicals	28	13	52	34	0	14	14	8	59	9	24	610	300	NY–NJ;Pa;Cg
Rubber and plastic	30	0	30	69	0	1	36	1	53	25	21	660	460	Cg;Cl;Dt
Primary metal products	33	12	53	32	0	6	5	5	64	14	17	380	330	Cg;Cl;Dt
Transportation equipment	37	0	89	11	0	0	31	0	73	16	11	520	470	NY–NY;NY–NJ;Pa; Ba;Ci;Dt;At

CLEVELAND

Other

Paints	285	22	2	76	0	30	34	14	38	31	16	670	280	Dt
Steel works	331	0	44	56	0	31	11	12	50	14	24	360	250	Ci;Cg;Pi;Dt
Iron and steel castings	332	0	4	96	0	36	4	11	41	31	18	150	140	Bu;Dt
Nonferrous shapes	335	0	12	87	0	27	19	19	45	17	19	610	390	NY-NJ;Ci;Bu;Dt
Nonferrous castings	336	0	0	99	0	20	1	16	38	17	28	230	230	Ci;Cg;Bu;Dt
Misc. primary metal	339	0	42	58	0	13	6	4	70	8	18	590	200	Ci;Cg;Dt

Fabricated metal

products	34	0	18	76	0	19	16	10	46	18	25	480	360	Ci;Cg;Dt
Nonelectrical machinery	35	0	59	40	0	41	18	13	48	16	23	730	400	NY-NY;Ba;Cg;Pi;Dt
Electrical machinery	36	0	20	76	0	23	34	18	53	14	16	720	520	Ci;Cg;Dt;SF;LA
Motor vehicles	371	0	66	33	0	12	25	3	70	14	13	580	580	NY-NJ;SL,Cg;Dt

DETROIT

Other

Paints	285	0	7	87	0	31	21	24	34	24	18	420	340	Pa
Cutlery and tools	342	0	2	85	0	41	22	27	36	30	6	350	330	Ci;Cg;Cl,LA
Bolts and nuts	345	0	0	99	0	36	9	27	38	15	21	370	210	Pa;Ci;Cg;Cl
Misc. fabricated metal	349	0	25	74	0	17	21	12	28	2	58	320	390	Pa;SL
Nonelectrical machinery	35	0	27	71	1	31	39	15	33	29	23	550	460	Cl
Motor vehicles	371	1	61	38	0	16	35	11	59	17	14	590	470	NY-NJ;Pa;Ci;SL,LA

MILWAUKEE

First stage

Beverage industries	208	0	38	62	0	18	37	2	36	27	34	550	400	Cg;Dt

Other

Chemicals	28	0	43	56	0	22	38	3	59	23	15	370	140	Ci;Cg;LA;At
Iron and steel castings	332	0	11	89	0	53	6	12	43	30	15	220	230	Cg
Nonelectrical machinery	35	0	35	60	1	12	46	4	30	22	43	620	570	NY-NY;Cg
Electrical machinery	36	0	28	48	10	12	66	2	59	21	18	990	560	NY-NY;Cg;Dl
Instruments	38	0	1	84	2	24	54	0	60	28	12	640	530	Pa;Cg;Dt;At

Table 14-5. Characteristics of Manufacturers' Intercity Shipments (tonnage) from San Francisco, Minneapolis, Los Angeles, and Houston Metropolitan Complexes, 1963

Complex and Commodity by Resource-Use Stage	SIC Code	TRANSPORT MODE				MILES SHIPPED		Production Area		Other Met- ro.	Non- Met- ro.	DESTINATION Other Met- ro.	Non- Met- ro.	Centers Receiving 4 percent
		Wa- ter	Rail	Mo- tor	Air	Under 100	Over 499	Same	Other			Miles		
		Percentage												
SAN FRANCISCO														
First stage														
Canning	203	6	71	23	0	9	78	7	54	28	12	1,520	1,780	NY–NY;Pa;Cg;LA
Misc. food preparations	209	1	48	51	0	30	43	21	35	32	12	850	800	LA
Petroleum	291	73	4	21	0	62	24	22	4	43	30	160	330	
Other														
Paints	285	1	28	71	0	46	20	26	36	23	15	210	180	LA;Se
MINNEAPOLIS														
First stage														
Food and kindred products	20	0	57	43	0	21	18	4	26	14	56	320	190	Cg
Other														
Nonelectrical machinery	35	0	28	68	0	3	73	1	44	26	29	840	620	NY–NY

LOS ANGELES

Other													
Clay and glass	32	0	33	67	0	80	1	71	1	17	11	180	170
Fabricated metal													
products	34	0	6	93	0	52	10	48	19	20	14	430	220	SF
Electrical machinery	36	1	3	89	3	20	58	19	47	22	11	1,240	1,250	Cg;SF
Aircraft	372	0	4	73	18	47	41	47	25	23	5	1,140	1,320	Dl;Se

HOUSTON

First stage													
Petroleum	291	98	1	1	0	2	90	2	63	25	10	860	790	Bo;NY–NY;NY–NJ; Pa;Ba
Other													
Chemicals	28	43	43	11	0	28	53	27	36	13	24	670	460	NY–NY;NY–NJ;Pa
Structural metal	344	5	31	64	0	15	26	12	20	32	36	500	450	DI
Nonelectrical machinery	35	0	36	64	1	10	51	8	16	36	41	580	670	DI

chinery, whose products enter long-distance flows terminating in both metropolitan and non-metropolitan territory. The market area for the products of Houston manufacturers in the structural-metal and nonelectrical-machinery industries appears to be sizable, but continuous; outputs of the petro-chemical industries, however, are directed in large share to the distant metropolitan complexes centered on Atlantic ports.

To know in detail the markets tapped by key industries in the remaining metropolitan complex, Los Angeles, would be particularly useful; for it is there that by far the largest concentration of manufacturing activity outside the northeast is to be found. The clay-and-glass industry is clearly "local" in orientation; four-fifths the intercity shipments are destined for locations within a hundred miles of the plant. The market area for fabricated metal products is broader, but still confined to the far west; only a tenth of the output moves as far as 500 miles. A still broader market area is associated with the electrical-machinery and aircraft specialties. Los Angeles manufacturers in these lines of industry appear to no more than penetrate the market east of Chicago, but to distribute their products widely outside the northeastern manufacturing belt.

There is ample evidence of a territorial division of labor within the manufacturing sector that extends beyond the differentiation of metropolitan from nonmetropolitan territory. Not only do metropolitan centers differ with respect to their manufacturing specialties, but they provide, one for another, important markets for the outputs of the key industries. If the nation is conceived as a mosaic of metropolitan regions, then interregional trade flows are more vital than intraregional flows in the marketing of manufactured goods.

Chapter 15

SPECIALITIES IN THE COMMERCIAL SECTOR

Local banks, insurance offices, wholesale houses, retail outlets, bus and rail depots, a commercial airport, schools, law offices, and hotels are among the facilities taken for granted by residents of communities far smaller than those singled out for study here. Some of these facilities, by their nature, serve nonlocal clienteles as well as local residents or economic units; facilities for interurban transportation or the lodging of transients are cases in point. Banking and educational institutions or wholesalers and lawyers may restrict their activities to the local arena, or they may maintain regular relations with extralocal clienteles. In general, if workers engaged in some branch of commerce are substantially overrepresented in the local work force relative to a nationwide norm, there is reason to suspect that the "market" extends beyond the local community although the possibility remains that an unusually heavy demand for the service is generated locally.

From the detailed industrial classification used for purposes of work-force tabulations in the 1960 Census, a group of titles is taken to represent the "commercial" sector, including finance, trade, transportation, and communication. The number of local specialties, as indexed by a location quotient of 2.0 or more, are relatively few in the major centers and, with only an occasional exception, are in branches of activity for which an extra-local clientele is intuitively plausible.

New York stands apart from the other centers of the northeast with respect to the richness of the profile in the commercial sector (Table 15-1). The financial preeminence of the metropolis, so fully documented in earlier chapters, is reflected in a specialization in security and commodity brokerage and investment companies and

Table 15-1. Industrial Profiles in the Commercial Sector for Boston, New York, Baltimore, Cincinnati, St. Louis, Chicago, Pittsburgh, and Buffalo, 1960

	LOCATION QUOTIENT FOR							
Detailed Industry	Bo	NY	Ba	Ci	SL	Cg	Pi	Bu
Finance								
Security and commodity brokerage and investment companies	2.1	3.9	...[a]	...	...	...	...	...
Insurance	1.8[b]	1.5	...	...	...	...	...	...
Banking and credit agencies	...	1.5	...	...	...	...	...	...
Trade								
Wholesale—								
Dry goods and apparel	2.2	5.3	...	...	2.4	...	...	...
Drugs, chemicals, and allied products	...	2.0	...	...	...	...	...	...
Retail—								
Fuel and ice dealers	2.0	...	...	...	...	...	...	...
General merchandise	...	...	...	...	...	1.6	...	...
Transportation								
Water transportation	...	3.0	2.9	...	...	...	...	...
Services incidental to transportation	...	2.9	...	...	...	...	...	...
Taxicab service	...	2.2	2.3	...	...	...	...	...
Street railways and bus lines	...	2.1	...	...	...	...	...	...
Railroads and railway express service	...	...	...	...	1.8	1.7	1.7	1.5
Communication								
Telegraph (wire and radio)	...	2.5	...	2.3	...	...	...	...

a. ... indicates location quotient less than 2.0, or less than 0.015 percent of national work force in industry.
b. Italic figures indicate at least 1.0 percent of local work force in industry.

a noteworthy overrepresentation of workers in the insurance industry and in banking and credit agencies. Specializations in the wholesaling of dry goods and apparel and of drugs, chemicals, and allied products distinguish the New York profile in trade. Transportation specialties include water transportation and services incidental to transportation, along with taxicab service and street railways and bus lines. A final specialty falls in the field of communication, namely, telegraph (wire and radio).

Boston alone shares with New York specializations in finance and in trade, but specialties in the transportation and communication sectors are absent from the Boston profile. Railroading is the sole specialty of Chicago, Pittsburgh, and Buffalo. St. Louis shares with these centers an overrepresentation of workers in the railroad industry and has, in addition, a specialty in the wholesaling of dry goods and apparel. Cincinnati has a specialty only in the communication field; both specialties of Baltimore are in transportation industries. The industrial structures of Philadelphia, Cleveland, Detroit, and Milwaukee reveal no overrepresentation of any commercial activity that might signify extralocal commercial functions.

This is not to say that sizable concentrations of commercial activity do not occur in the northeastern centers; indeed, such concentrations do exist. Chicago, for example, is found to have a location quotient only slightly greater than unity for security and commodity brokerage and investment companies; yet the Chicago concentration accounts for about five per cent of the national work force engaged in that branch of commerce. Assuming that strong extralocal functions will be reflected in a high location quotient for the industry, however, the commercial institutions of all major centers of the northeast save New York are oriented to a largely local market.

A rich profile in the commercial sector is more characteristic of major centers in the south or west than of the northeastern centers. Only two of the twelve northeastern centers had more than two commercial specialties; only two of the ten centers to the south or west had fewer than two commercial specialties. New York has been shown to be a clear exception to this generalization, however; and among the centers to the south or west, it is Los Angeles that proves to be the striking exception. No commercial specialty was detected in the Los Angeles industrial structure.

The roster of commercial specialties in the old ports of New Orleans and San Francisco is somewhat reminiscent of the commercial profile in the metropolitan complex centered on the port of New York. Strong specializations in water transportation and in the services incidental to transportation appear in the profiles of New Orleans and San Francisco. San Francisco has developed a supplementary specialty in air transportation, and noteworthy overrepresentations of workers in the insurance industry and in banking and credit agencies are to be observed in the profile. Telegraph communication and the wholesaling of electrical goods, hardware, and plumbing equipment and of food and related products round out the roster of specialties in New Orleans (Table 15-2).

Minneapolis and Kansas City, the rail gateways to the mid-continent plains, have retained railroading as a prominent feature of the industrial profile. Each center has gained air transportation as a supplementary specialty, and Kansas City specializes in still another new transport mode, petroleum and gasoline pipelines. Both centers are active in wholesaling, with the specialties including electrical goods, hardware, and plumbing equipment, motor vehicles and equipment, and farm products in Minneapolis and drugs, chemicals, and allied products and motor vehicles and equipment in Kansas City. Moreover, their retailing activities in the general merchandise line appear to extend beyond the local market. An insurance specialty is present in both centers, and Minneapolis has also a specialty in security and commodity brokerage and investment companies. The final specialty is telegraph communication which is observed only in the Kansas City profile.

The new center of Dallas resembles the older gateway centers with an interesting exception; its sole specialty in the transportation field is air transport. The wholesaling specialties of Dallas include dry goods and apparel, electrical goods, hardware, and plumbing equipment, motor vehicles and equipment, and machinery, equipment, and supplies. Specialties in insurance and in telegraph communication also appear in the Dallas profile.

The commercial sector of Houston is oriented to the petroleum industry. In fact, the extraction of crude petroleum and natural gas in the Houston area is the most pronounced extractive specialty to be observed among the major centers. (The other extractive specialties are petroleum and gas in Dallas and New Orleans, coal

Table 15-2. Industrial Profiles in the Commercial Sector for New Orleans, San Francisco, Washington, Minneapolis, Kansas City, Houston, Dallas, Seattle, and Miami, 1960

Detailed Industry	LOCATION QUOTIENT FOR								
	NO	SF	Wa	Mp	KC	Ho	Dl	Se	Mi
Finance									
Security and commodity brokerage and investment companies	...[a]	...	...	2.0	...	...	...	...	...
Insurance	...	*1.7*[b]	...	*1.6*	*1.5*	...	2.3	...	...
Banking and credit agencies	...	*1.5*	...	...	...	...	...	...	...
Trade									
Wholesale—									
Dry goods and apparel	...	...	...	...	...	...	2.5	...	...
Drugs, chemicals, and allied products	...	...	...	...	2.4	...	...	...	...
Electrical goods, hardware, and plumbing equipment	2.2	...	...	2.4	...	...	2.2	...	...
Food and related products	2.0	...	...	...	...	...	...	...	...
Motor vehicles and equipment	...	...	...	2.2	2.5	...	2.7	...	...
Farm products—raw materials	...	...	...	2.1	...	...	...	...	...
Petroleum products	...	...	...	...	...	2.9	...	...	...
Machinery, equipment, and supplies	...	...	...	...	...	*2.8*	2.6	...	...
Retail—									
General merchandise	...	...	...	*1.7*	*1.5*	...	...	...	...
Transportation									
Water transportation	*13.1*	*4.2*	...	...	...	*4.3*	...	*4.4*	...
Services incidental to transportation	2.6	2.6	...	...	...	...	...	...	...
Taxicab service	2.7	...	2.7	...	...	...	...	...	...
Railroads and railway express service	...	...	...	*1.9*	*2.3*	...	...	...	...
Air transportation	...	*3.5*	2.7	*3.1*	*4.5*	...	*4.1*	3.0	*17.0*
Petroleum and gasoline pipe lines	...	...	...	...	4.1	8.5	...	...	...
Communication									
Telegraph (wire and radio)	2.0	...	...	...	2.0	...	3.2	...	...

a. ... indicates location quotient less than 2.0, or less than 0.015 percent of national work force in industry.

b. Italic figures indicate at least 1.0 percent of local work force in industry.

mining in Pittsburgh, and fisheries in New Orleans and Seattle.)
The wholesaling of petroleum products and of machinery, equipment, and supplies are key activities in Houston, along with water
transportation and petroleum and gasoline pipe lines.

The specialties of the other new centers are to be found in the
transportation field: water and air transportation in Seattle; and air
transportation alone in Miami. The profile of the remaining center,
Washington, is distinguished by specialties in air transportation and
taxicab service.

Systematic examination of the correspondence between a center's age and the age of its commercial specialties is impossible, for
we cannot assign "ages" to the full range of specialties. We can,
however, distinguish old from new transport modes and old from
new lines of products handled by wholesalers.

The oldest of the transport modes is water transportation; somewhat newer is rail transportation; newer still are the pipe lines and
airways. The fifteen major centers with a transportation specialty
are arranged below by the type(s) of transport mode that distinguishes them.

	Alone	*With pipe or air*
Water	NY, Ba, NO	SF, Ho, Se
Rail	SL, Cg, Pi, Bu	Mp, KC
Air	Wa, Dl, Mi	

In no instance is a center's sole transportation specialty in a transport mode which antedates the center's rise to national prominence
with the exception of Washington, a special case in terms of economic function.

A somewhat similar relation of age of center and age of specialty
appears in the wholesaling field. Wholesaling in the dry-goods,
drugs, food, and farm-products lines probably would be conceded
to be older than the wholesaling of machinery, electrical goods, motor vehicles, and petroleum products. A grouping of centers with a
wholesaling specialty by the age of specialty is shown below.

	Alone	*With new*
Old	Bo, NY, SL	NO, Mp, KC
New	Ho, Dl	

A few other types of service activity appear as local specialties
in a major center. Some are clearly an outgrowth of local commerce

which itself is oriented to an extralocal clientele. Other "service" specialties seem to represent an extension of the center's preeminence in another line of activity.

A "service" specialty appears in the profiles of six centers of the northeast. The New York cluster includes firms providing miscellaneous professional and business services, advertising agencies, and theaters and motion pictures. In Boston, it is local public administration, the miscellaneous professional services, private educational institutions, and engineering and architectural firms that form the specialty cluster. Miscellaneous professional services and advertising agencies are prominent in the Chicago profile, and workers in advertising agencies are overrepresented in the Detroit work force. A public-administration specialty is present in two other northeastern centers: local administration in Milwaukee; federal administration in Baltimore, where it almost certainly represents spillover from neighboring Washington. (See Table 15-3.)

Table 15-3. Industrial Profiles in the Service Sector for Boston, New York, Baltimore, Chicago, Detroit, and Milwaukee, 1960

Detailed Industry	LOCATION QUOTIENT FOR					
	Bo	NY	Ba	Cg	Dt	Ml
Administration						
Local public	1.5[b]	...[a]	...	...	...	1.5
Federal public	...	...	1.7	...	...	...
Other						
Educational services: private	2.6	...	...	...	...	...
Engineering and architectural services	2.5	...	...	...	...	...
Misc. professional and related services	2.2	2.1	...	2.1	...	...
Advertising	...	3.2	...	2.1	2.1	...
Theaters and motion pictures	...	2.1	...	...	...	...
Misc. business services	...	2.0	...	...	...	...

a. ... indicates location quotient less than 2.0, or less than 0.015 percent of national work force in industry.

b. Italic figures indicate at least 1.0 percent of local work force in industry.

The profiles of seven southern or western centers include a "service" specialty although distinctive clusters appear only in the profiles of Washington and Miami. The Washington specialties include not only federal public administration, but also miscellaneous professional and business services, nonprofit membership organizations, and legal and medical services. The Miami cluster includes legal and real-estate services, hotels, and miscellaneous entertainment and recreation services, as well as local public administration. The overrepresentation of federal public administration workers in San Francisco may reflect the city's role as metropolis for the far west when the western territory was integrated into the eastern-based metropolitan organization; combined with the administrative specialty is a specialty in miscellaneous professional and business services. The specialties of Los Angeles are business services and the theaters and motion pictures. Warehousing and storage are specialties of both New Orleans and Houston; workers in the water-supply industry are overrepresented in New Orleans; and Houston shares with Dallas a specialization in "gas and steam supply systems." These last and rather unusual specialties may reflect uncommon features of the site of a major center. (See Table 15-4.)

The overall patterning of differences among centers in commercial and "other service" specialties manifests an interesting feature. New York, by far the nation's largest metropolitan complex, exhibits a rich profile in the commercial and service sectors. The centers which have, at some point in the nation's history, been contenders for second rank in the city-size distribution, Philadelphia, Chicago, Detroit, and Los Angeles, exhibit profiles that are almost devoid of such specialties, however. The profile of New York is most nearly approximated, though never matched, by the profiles of centers occupying substantially lower ranks in the city-size distribution. Although a cluster of specialties in the commercial and service sectors may be a hallmark of the metropolis in the strict sense, such a cluster is neither necessary nor sufficient for the emergence of a great urban complex. Indeed, the centers offering the strongest challenge to the nation's "first city" with respect to population size or scale of economic activity are those whose functions complement rather than replicate the repertoire of the national metropolis.

Although a cluster of commercial or service specialties in a center's industrial profile is almost certain to signal strong extralocal functions, the absence of such a cluster does not necessarily mean that the commercial and service sectors perform only local mainte-

Table 15-4. Industrial Profiles in the Service Sector for New Orleans, San Francisco, Washington, Los Angeles, Houston, Dallas, and Miami, 1960

	LOCATION QUOTIENT FOR						
Detailed Industry	NO	SF	Wa	LA	Ho	Dl	Mi
Administration							
Local public	...[a]	...	...	...	...	...	*1.6*[b]
Federal public	...	*1.8*	*12.5*	...	...	...	...
Other							
Educational services: private	*1.7*	...	...	...	...	...	...
Misc. professional and related services	...	3.1	3.5	...	...	...	...
Theaters and motion pictures	...	...	...	*4.2*	...	...	...
Misc. business services	...	2.0	*2.1*	*2.2*	...	...	...
Water supply	2.7	...	...	...	...	...	...
Warehousing and storage	2.2	...	...	...	2.6	...	...
Private households	*1.5*	...	...	...	...	...	*1.6*
Nonprofit membership organizations	...	...	3.8	...	...	...	...
Legal services	...	...	*2.1*	...	...	...	*2.2*
Medical and other health services, except hospitals	...	...	*1.5*	...	...	...	...
Gas and steam supply systems	...	...	...	...	4.2	2.3	...
Hotels and lodging places	...	...	...	...	...	...	*5.0*
Real estate	...	...	...	...	...	...	*2.5*
Misc. entertainment and recreation services	...	...	...	...	...	...	2.5

a. ... indicates location quotient less than 2.0, or less than 0.015 percent of national work force in industry.

b. Italic figures indicate at least 1.0 percent of local work force in industry.

nance functions. Strength in a center's commercial or service sector can arise either from the development of a few specialties oriented to a near national market or the development of a broad range of activities oriented to a more restricted, yet extralocal territory.

As an alternative index of the strength of commercial activity, we rely here on the per capita value of sales by wholesalers located in the center. The five centers in which per capita sales took on the highest values in 1963 are places in which a cluster of commercial specialties has been identified, namely, New York, San Francisco, Minneapolis, Kansas City, and Dallas. Wholesaling is no stronger in New Orleans and Houston, the other centers with a cluster of commercial specialties, than in Boston, Cincinnati, St. Louis, Chicago, and Cleveland, however.

Strong service sectors as indexed by a relatively high per capita value of receipts by local service establishments, are found in several centers where no cluster of service specialties appeared in the industrial profile. Distinguishing clusters of service specialties were observed not only in New York and Miami, where per capita receipts are at a maximum, but also in Washington and Boston, centers ranking eighth and tenth with respect to per capita receipts. The intervening ranks with respect to per capita receipts are occupied by Chicago and Detroit, San Francisco, Los Angeles, Minneapolis, and Dallas.

A schematic display of centers corresponding to the verbal description above may serve as a useful overview. Centers in which a cluster of specialties in the commercial sector was identified are denoted by a C; those in which a cluster of specialties in the service sector was identified are denoted by an S. Centers where the per capita wholesale sales falls within the range observed for C centers are denoted as having "trade strength." Centers where per capita service receipts fall within the range observed for S centers are denoted as having "service strength."

TRADE AND SERVICE STRENGTH

New York–CS

San Francisco–C	
Minneapolis–C	Boston–S
Dallas–C	

Chicago

TRADE STRENGTH	SERVICE STRENGTH
New Orleans–C	
Kansas City–C	Washington–S
Houston–C	Miami–S
Cincinnati	Detroit
St. Louis	Los Angeles
Cleveland	

NEITHER

Philadelphia	Buffalo
Baltimore	Milwaukee
Pittsburgh	Seattle

The position of New York as the nation's "first city" is established once more, for it is the only center in which strength in both the trade and service sectors is observed in conjunction with specialties clusters in both the commercial and service sectors. A coincidence of strength in the service and commercial sectors is observed in only five other major centers: San Francisco, Minneapolis, and Dallas, where a commercial-specialties cluster appears in the industrial profile; Boston, where a service-specialties cluster appears in the industrial profile; and Chicago, where neither type of cluster is present. One is tempted to infer that these centers, scattered across the nation at very considerable distances from one another, are the strategic points at which the distribution of goods and services to ultimate consumers is organized.

Chapter 16

THE RISE OF A TEXAS METROPOLIS

Although the financial network now centered on Dallas extends well beyond the limits of the state of Texas, the financial pre-eminence of the city rests in part on its selection as a correspondent-bank location by four-fifths of the Texas banks. Not until after 1900, however, did Texas bankers prefer Dallas over all other Texas cities, let alone over such out-state financial centers as St. Louis, Kansas City, and New Orleans or still more distant Chicago and New York.

We can establish a baseline reading on the correspondent preferences of Texas banks at a time when no city within the state had as many as 25,000 residents and when some 124 in-state banks met the needs of local residents and provided the links into the national money market. Moreover, the baseline antedates the extraction of petroleum in mid-continent, which created new wealth within the state and became a competitor of timber-cutting, cotton-farming, and cattle-raising for indigenous capital.

As late as 1800, Galveston and San Antonio were the contenders for first rank among the Texas cities in population size. Galveston with 22,000 residents held a lead of 1,000 persons over San Antonio. Galveston's lead as an in-state financial center was more secure. A sixth of the Texas banks maintained a Galveston correspondent in contrast to the 1 percent choosing any other Texas city. Most Texas bankers, however, relied on banks in more important out-state centers, notably, New York and St. Louis, for correspondent services.

By 1900 San Antonio with 53,000 residents had established a 15,000-person lead over Galveston. Moreover, both Dallas and Houston had shifted ahead of Galveston in rank with respect to population size. Still Galveston remained the most popular in-state location for a correspondent bank, again the choice of a sixth of the Texas bankers. The contenders for second rank as a Texas financial center were Dallas and Houston, each the choice of about an eighth of the Texas bankers. Although Fort Worth as well as Dallas and Houston had expanded their correspondent-banking roles since 1880, the dominant positions of New York and St. Louis remained unchanged. More than nine-tenths of the Texas banks continued to maintain a New York correspondent; three-fifths maintained a St. Louis correspondent. The out-state centers of Chicago, New Orleans, and Kansas City has expanded their correspondent roles in Texas, and by 1900 they were chosen nearly as often as any Texas city was chosen. (See Tables 16-1 and 16-2.)

Table 16-1. Population of Selected Cities in Texas, 1860 to 1965 (in thousands)

| | CITY PROPER | | | | | | | METROPOLITAN AREA | |
Place	1860	1880	1900	1920	1940	1950	1960	1960	1965
San Antonio	8	21	53	161	254	408	588	687	807
Galveston	7	22	38	44	61	67	67	140	n.a.
Houston	5	17	45	138	385	596	938	1,243	1,695
Austin	3	11	22	35	88	132	187	212	n.a.
Dallas	...	10	43	159	295	434	680	1,084	1,289
Fort Worth	...	7	27	106	178	279	356	573	627
El Paso	...	1	16	78	97	130	277	314	n.a.

Note: n.a. indicates estimate not available.

Some radical shifts in the positions of both in-state and out-state centers as correspondents for Texas banks have occurred since 1900. The importance of distant centers has declined sharply, most notably perhaps in the case of St. Louis. At the same time, there has been an extraordinary expansion in the correspondent activities of many Texas cities, including Dallas of course.

Table 16-2. Percentage of Banks in Texas with a Principal Correspondent in Selected Cities in 1881, 1900, 1940, 1950, and 1965

Location of Correspondent	1881	1900	1940	1950	1965	Location of Correspondent	1881	1900	1940	1950	1965
Out-State						In-State—cont.					
New York	95.2	94.0	47.9	46.7	39.2	San Antonio	0.8	3.6	16.2	17.1	18.0
St. Louis	58.9	61.0	8.1	6.0	3.7	Austin	0.0	3.3	5.0	7.2	14.2
New Orleans	7.3	15.7	2.1	3.2	1.7	Waco	0.0	3.1	4.0	3.6	5.1
Chicago	4.0	14.5	9.3	8.9	6.9	Amarillo	0.0	0.2	5.3	4.9	6.7
Kansas City	0.0	18.8	14.2	12.7	7.4	Wichita Falls	0.0	0.2	3.9	3.0	4.2
San Francisco	0.0	0.2	3.2	3.1	1.9	San Angelo	0.0	0.2	0.7	0.6	1.1
Los Angeles	0.0	0.0	2.2	2.8	3.6	Corpus Christi	0.0	0.0	2.1	2.1	2.7
Atlanta, Ga.	0.0	0.0	0.0	1.2	1.1	El Paso	0.0	0.0	1.4	2.6	3.1
All other	0.0	1.0	8.0	8.1	7.0	Lubbock	0.0	0.0	1.4	2.5	6.4
In-State						Beaumont-Port Arthur	0.0	0.0	1.0	0.8	2.9
Galveston	16.1	16.6	6.9	3.2	2.5	Laredo	0.0	0.0	0.2	0.1	0.2
Fort Worth	1.6	7.7	34.4	31.8	31.0	All other	0.8	7.2	17.7	16.5	24.6
Dallas	0.8	12.8	66.5	72.8	78.4						
Houston	0.8	11.6	43.6	44.2	44.0						

Galveston's position in the ranking of Texas cities both with respect to population size and financial importance has slipped. Its population has less than doubled since 1900, and it is a less popular location for a correspondent bank now than at any time in the past. A turning point in the city's fortunes may have been the devastating hurricane and tidal wave which hit in the fall of 1900. Growth was further stunted when an inland ship channel was completed which made Houston fully competitive as a Gulf coast shipping point.

By 1920, San Antonio, Houston, Dallas, and Fort Worth had become rather sharply differentiated from the other cities in Texas in terms of population size; and they have remained the State's largest centers. In financial importance, Dallas has maintained a commanding lead over second-placed Houston which, in turn, has maintained a rather narrow lead over third-placed Fort Worth. San Antonio has not emerged as a serious challenger to Dallas, Houston, or Fort Worth in the financial sphere.

New York's banking role in Texas has diminished greatly; the proportion of Texas institutions maintaining a New York correspondent fell from over nine-tenths in 1900 to half in 1940 and two-fifths in 1965. St. Louis, however, has been more or less wiped out as a financial center for Texas; once the choice of three-fifths of the Texas bankers, it was chosen by only 4 percent in 1965. The positions of Chicago and Kansas City also have declined, although less sharply than that of St. Louis, and New Orleans has essentially been eliminated as a correspondent-banking center for Texas institutions.

A somewhat different view of the competition among centers can be gained by mapping the sections of the state in which a given center was frequently chosen as a correspondent-bank location. First, it must be recognized that many counties in Texas, most often in the western part of the state, had no bank in 1900. (See Figure 16-1. The mapping is based on the county delineations as of 1960 although some county lines have changed since 1900.) Competition between two centers located outside the state is traced spatially before the rivalry between Texas centers is displayed.

The influence of spatial factors on the competitive positions of centers located at some distance can be seen by examining the correspondent hinterlands of St. Louis and Kansas City. Kansas City lies some 500 miles due north of the easternmost point in

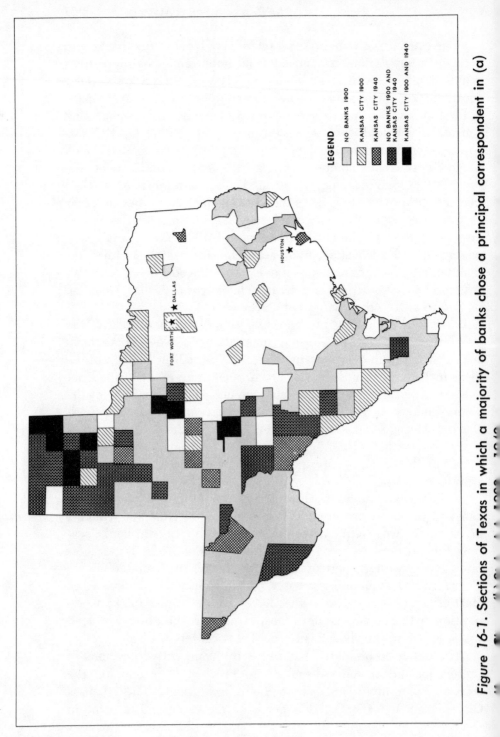

LEGEND

	NO BANKS 1900
	KANSAS CITY 1900
	KANSAS CITY 1940
	NO BANKS 1900 AND KANSAS CITY 1940
	KANSAS CITY 1900 AND 1940

Figure 16-1. Sections of Texas in which a majority of banks chose a principal correspondent in (a)

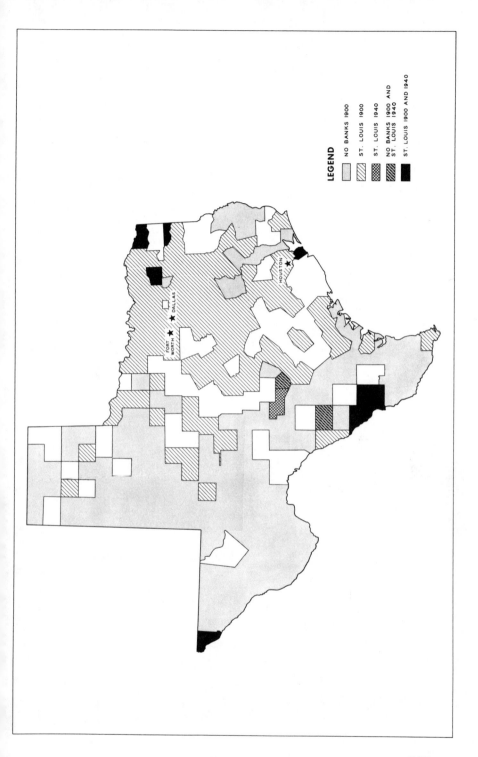

LEGEND

NO BANKS 1900

ST. LOUIS 1900

ST. LOUIS 1940

NO BANKS 1900 AND
ST. LOUIS 1940

ST. LOUIS 1900 AND 1940

HOUSTON

DALLAS

FORT
WORTH

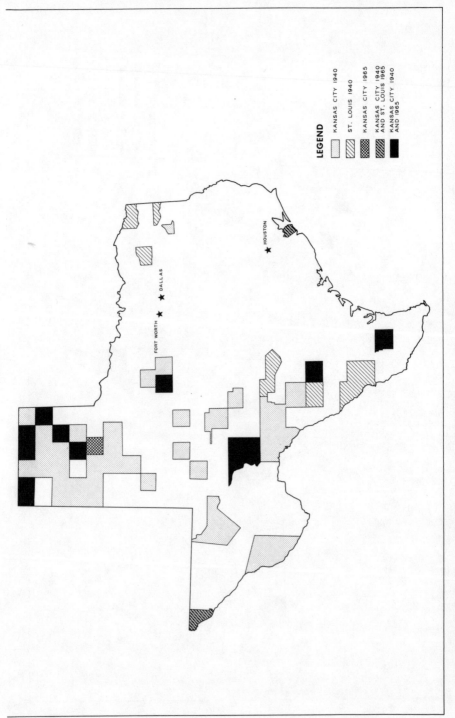

LEGEND

KANSAS CITY 1940

ST. LOUIS 1940

KANSAS CITY 1965

KANSAS CITY 1940
AND ST. LOUIS 1965

KANSAS CITY 1940
AND 1965

Figure 16-2. Sections of Texas in which a majority of banks chose a principal correspondent in Kan-

Texas, and St. Louis lies some 250 miles due east of Kansas City. Accordingly, Kansas City enjoys something of a distance advantage to banks in Texas relative to St. Louis. In eastern Texas, the distance advantage of Kansas City is typically something under 150 miles; but it becomes greater to the west and in the Panhandle.

In 1900 St. Louis was chosen as a correspondent-bank location by three-fifths of the Texas banks; Kansas City was the choice of more than a sixth. St. Louis then had an extensive hinterland which stretched through the eastern section of the state. Toward the west, and especially in the Panhandle, the position of Kansas City relative to St. Louis became stronger. The hinterland controlled by these Missouri cities was transitory, however. By 1940 neither center retained east Texas in its correspondent-banking hinterland, and parts of western Texas had been lost as well. St. Louis had become the choice of less than a tenth of the Texas banks, but Kansas City remained the choice of a seventh of the Texas banks. This suggests that Kansas City was more successful than St. Louis in adding new territory to its hinterland after 1900; and, indeed, it was. Its role as a correspondent-banking center for banks in west Texas and the Panhandle expanded substantially (Figure 16-2).

After 1940 Kansas City was unable to block the development of correspondent centers within the state, and its position had deteriorated rather badly by 1965. In only eleven counties did as many as half the banks select Kansas City as a correspondent location, and only one of these counties represented territory gained since 1940. St. Louis no longer was the choice of a majority of the banks in any Texas county. It is of interest that the losses of the Missouri cities first occurred in the eastern part of the state where the larger Texas cities were located. To the west, Kansas City was able to compete more effectively than the in-state centers of Dallas, Houston, and Fort Worth as well as the out-state center of St. Louis. In a sense, Kansas City remained an important center longer than St. Louis because it was more protected from competition with in-state correspondent centers.

Galveston and Houston are located within some 50 miles of one another. At the turn of the century, they were competitors in the southeastern part of the state although neither was sufficiently developed as a financial center to have statewide importance. Dallas and Fort Worth are neighboring cities located some 200

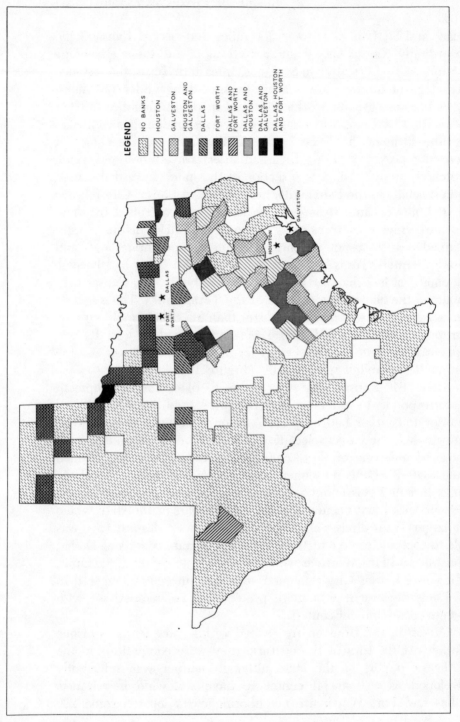

LEGEND

- NO BANKS
- HOUSTON
- GALVESTON
- HOUSTON AND GALVESTON
- DALLAS
- FORT WORTH
- DALLAS AND FORT WORTH
- DALLAS AND HOUSTON
- DALLAS AND GALVESTON
- DALLAS, HOUSTON AND FORT WORTH

Figure 16-3. Sections of Texas in which a majority of banks chose a principal correspondent in Dallas, Fort Worth, Houston, or Galveston in 1890.

miles to the north of Houston. Fort Worth's strength was drawn from the territory to its west, including the Panhandle. Dallas, by contrast, was much more successful to the east and fared better in the area to the south. (See Figure 16-3.)

After the turn of the century, the role of Galveston as a correspondent center began to deteriorate. By 1930 the rivalry among in-state centers for the Texas hinterland was restricted to Fort Worth, Dallas, and Houston, each of which had carved out a fairly clear-cut hinterland. The advantage of Dallas, the state's most important correspondent center, did not stem from exclusive control of an unusually large section of the state. Fort Worth and Houston monopolized sections of Texas as extensive as the section controlled by Dallas. There were, however, few parts of the state in which Fort Worth and Houston shared a hinterland, in contrast to the many sections in which Dallas shared a hinterland with either Fort Worth or Houston. The advantage of Dallas was the extent of the territory in which it played an active correspondent role, alone or in conjunction with another center. (See Figure 16-4.)

During the 1930s Dallas expanded its correspondent position in the areas to its west, including the Panhandle. Although Fort Worth's hinterland also grew somewhat, there were relatively few counties in 1940 that were served by Fort Worth but not by Dallas. Moreover, the Dallas hinterland in southeastern Texas expanded considerably, reaching nearly to the counties bordering the Gulf. A pattern developed in the rivalry between Dallas and Houston analogous to the competition between Dallas and Fort Worth. Houston's hinterland expanded, particularly in the areas to the west; but the expansion of Dallas was much greater and resulted in a drastic decline in the counties monopolized by Houston. There had become by 1940 a broad belt of counties between these two centers which turned to both cities for correspondent-banking services. Dallas was unable to replace either Houston or Fort Worth as a correspondent center, but it was able to capture a share of the correspondent activity in their hinterlands (see Figure 16-5).

The changes in hinterlands since 1940 have been relatively minor and are somewhat inconclusive. With respect to the Dallas-Fort Worth rivalry, it is clear that in 1965 Dallas banks are active everywhere that Fort Worth banks are active and, in addition, are active in other sections of the state. The number of counties served

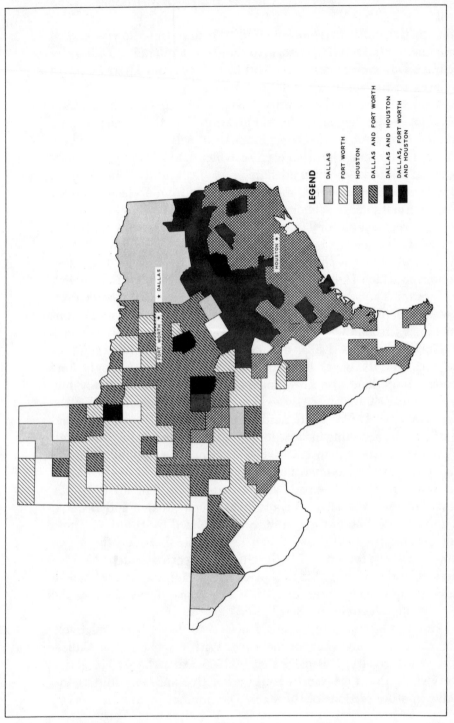

LEGEND

DALLAS

FORT WORTH

HOUSTON

DALLAS AND FORT WORTH

DALLAS AND HOUSTON

DALLAS, FORT WORTH
AND HOUSTON

Figure 16-4. Sections of Texas in which a majority of banks chose a principal correspondent in Dallas, Fort Worth, or Houston in 1920.

exclusively by Fort Worth has declined, and the expansion of Dallas banks as correspondents in west Texas has been unmatched by Fort Worth banks. Whether over the next few decades Fort Worth's hinterland will wither, or whether a state of equilibrium has been reached is difficult to say. A notable change in the competition between Dallas and Houston has been further decline in the hinterland monopolized by Houston, an area now consisting of only a dozen or so counties scattered over southeast Texas. Trend in Houston's influence in central Texas is more difficult to evaluate. A belt of counties running from the Mexican border toward Dallas now are served exclusively by that center; a number of counties in western Texas were being served by Houston as well as by Dallas, however (see Figure 16-6).

Two spatial aspects of the competition of Dallas with Fort Worth and Houston are rather surprising. First, Fort Worth's hinterland extends westward to the edge of the state and through nearly all of the Panhandle, but it is completely blocked to the east. Second, Houston's hinterland runs northward to a point fairly close to Dallas.

At least part of the reason that Fort Worth banks are most successful in obtaining correspondent business in west Texas lies in the economic activities pursued in that section of the state and the kind of specialization in banking functions found in Fort Worth and neighboring Dallas. A letter from Mr. Lewis H. Bond, President of The Fort Worth National Bank, is illuminating on the point.

> First of all, Dallas is the acknowledged financial center of the southwest. . . . The Federal Reserve Bank for the Eleventh District is also located in Dallas. Since banks ordinarily like to correspond with the largest regional bank available, feeling that the larger the bank, the more benefits to be obtained from the relationship, the question comes down to this: why don't the Dallas banks have the same dominant position in West Texas as East Texas? Or, conversely, how can we compete as effectively as we do in the western part of the state?
>
> The answer to this, I think, lies in the long-established ties which Fort Worth has to the West—ties which were established when the economy of that area was based largely on cattle and agriculture, and Fort Worth was the principal market for these products. As a consequence of this, the Fort Worth banks specialized early

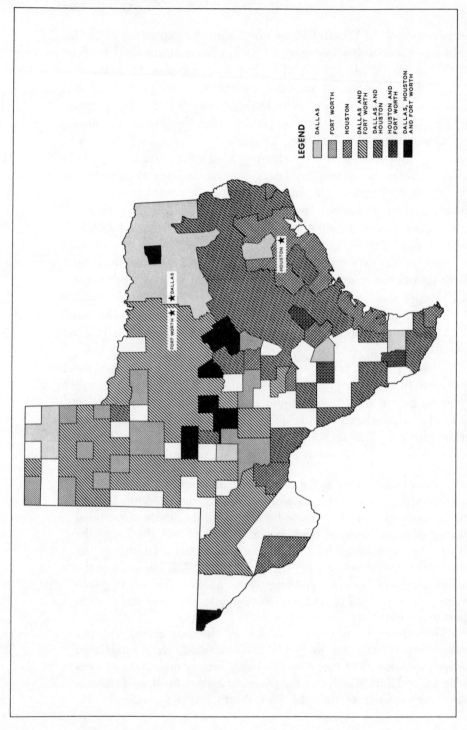

LEGEND

DALLAS

FORT WORTH

HOUSTON

DALLAS AND
FORT WORTH

DALLAS AND
HOUSTON

HOUSTON AND
FORT WORTH

DALLAS, HOUSTON
AND FORT WORTH

Figure 16-5. Sections of Texas in which a majority of banks chose a principal correspondent in Dallas, Fort Worth, or Houston in 1940.

in livestock and agricultural financing and established connections which have been maintained and strengthened even though conditions have changed considerably.

The Dallas banks offset this disadvantage somewhat in the early 1930's when they became active in financing oil and gas development. Banks in Fort Worth were much less aggressive in this field until some fifteen or twenty years later and missed a fine opportunity to capitalize on this growing industry and tie West Texas even closer to this city. Our banks are now competitive in this field, but the late start has made a difference.

. . . Dallas and Fort Worth banks perform very similar functions for their correspondents, except that the Fort Worth banks provide ranching and agricultural know-how probably not available in Dallas.

But many of the banks that call on us for most correspondent services still like to have an account with one of the larger banks in the larger city. I wish that we could convince them it is not necessary.

The extension of Houston's correspondent services into an area so close to Dallas seems less readily explained. One important consideration is the Federal Reserve Branch District delineated for Houston; the District runs along the Gulf coast and northward toward Dallas. This organizational factor itself reflects certain economic relations in the same way that Fort Worth's activity as a meat packer provides a linkage to the cattle-raising country of west Texas. The hinterlands of Galveston and Houston extended northward into territory proximate to Dallas and Fort Worth prior to the establishment of the Federal Reserve System (Figure 16-3). It seems likely that the Gulf ports initially were linked with these sections of east Texas as outlets for the timber-cutting and cotton-farming industries. The tie was reinforced with the development of the petroleum industry.

A final issue is whether the selection of Dallas as a Federal Reserve City in 1914 was an outcome or a precondition of the city's financial importance. It will be recalled that in 1900 Galveston, Dallas, Houston, and Fort Worth were nearly matched in terms of their correspondent role among Texas bankers. At the time the Federal Reserve Districts were being delimited, a poll of Texas banks indicated a strong preference for headquartering the reserve bank in Dallas, however. Dallas received 212 first-choice votes,

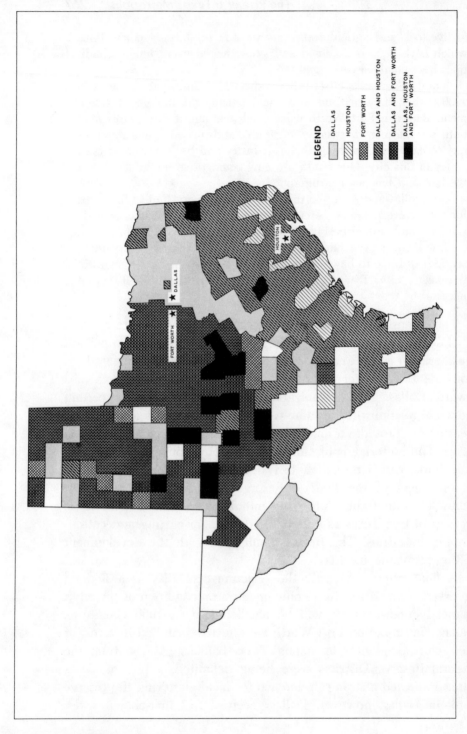

Figure 16-6. Sections of Texas in which a majority of banks chose a principal correspondent in Dallas, Fort Worth, or Houston in 1965.

LEGEND

DALLAS

HOUSTON

FORT WORTH

DALLAS AND HOUSTON

DALLAS AND FORT WORTH

DALLAS, HOUSTON
AND FORT WORTH

compared with 97 for Houston, 84 for Fort Worth, and still fewer for Galveston (United States Senate, 1914: 355). Nevertheless, some 27 percent of the Texas banks sampled in a 1914 Directory maintained a principal correspondent in Houston, as compared with 23 percent in Dallas and 22 percent in Fort Worth. Whatever the long-run effect of the creation of a Federal Reserve Bank District headquartered in Dallas may have been, there was no immediate shift in the correspondent choices of Texas banks among the Texas cities. As late as 1930 Houston trailed Dallas as a correspondent center by only a slight margin (39 versus 42 percent), and Fort Worth remained a strong competitor (with 31 percent). Not until the 1930s did Dallas consolidate its position as the state's leading correspondent center. In that decade during which Dallas expanded its services to reach two-thirds of the Texas banks, Houston and Fort Worth remained static. The timing suggests that it was not status in the Federal Reserve System alone which gave Dallas primacy although this status may have worked to the advantage of Dallas banks as they established ties with the growing petroleum industry.

In theory, because banks have been increasing the number of principal correspondents they select, a city can carve out an important correspondent role in a territory without driving out correspondent banks based in another city. In practice, however, the expansion of a city's correspondent-banking services in a territory usually is accompanied by at least a partial decline in the importance of older centers. In establishing itself as a major center in the nation's financial network, Dallas appears first to have undercut the importance of in-state centers on which networks of limited extent were centered and then to have displaced older out-state centers of long-standing financial importance which had included Texas as part of a far ranging hinterland.

Chapter 17

THE COMPETITION FOR BANK CORRESPONDENTS IN FLORIDA

Florida in 1900 was almost undeveloped territory given the growth which the state has since experienced. Its population was then half a million, compared with nearly five million in 1960. During the six decades in which Florida's population grew nearly ten-fold, the United States population little more than doubled.

The largest city in Florida in 1900 was Jacksonville with 28,000 inhabitants; the other leading centers were Pensacola, Key West, and Tampa. Miami, which was to become the state's largest city, had a population totaling only 1,681. Although Jacksonville's growth during this century has been considerable, particularly in the opening decades, its expansion has not matched that of other cities in the State. Miami began to narrow Jacksonville's population advantage in the 1920s, and by 1940 each city had a population of slightly less than 175,000 inhabitants. The 1960 population in metropolitan Miami numbered close to a million, twice the number in the Jacksonville metropolitan area. The Tampa-St. Petersburg metropolitan area had not gained residents as rapidly as the Miami area, but its population exceeded Jacksonville's population by more than 300,000 in 1960.

What changes have occurred in the locations of correspondents selected by banks in Florida since 1900? New York was selected as the location of a principal correspondent by all but one of Florida's fifty-eight banks in 1900. Jacksonville occupied second place, clearly subordinate to New York, but secure from any rivals. Jacksonville banks served as a principal correspondent for half the state's banking institutions. The closest in-state competitor was Tampa, a city selected by six neighboring banks as a principal correspondent. The out-state centers of Atlanta and Chicago were selected by three and two banks, respectively. Keeping in mind that more than half the counties in Florida had no bank in 1900, Jacksonville's correspondent hinterland was statewide, with the exception of that part of northwestern Florida closest to New Orleans and several counties near Tampa (Figure 17-1). It was not unreasonable to designate Jacksonville as a Federal Reserve Branch City of the Atlanta District in 1914.

The locations of bank correspondents selected by Florida institutions in the years 1900, 1940, 1950, and 1965 (Table 17-1) reflect three types of competition: first, the efforts of major national banking centers to obtain a position in Florida; second, the competition between banks in Florida cities to develop hinterlands within

Table 17-1. Percentage of Banks in Florida with a Principal Correspondent in Selected Cities in 1900, 1940, 1950, and 1965

Location of Correspondent	1900	1940	1950	1965	Location of Correspondent	1900	1940	1950	1965
Out-State					In-State—cont.				
New York	98.3	91.8	91.2	87.8	Miami	0.0	8.4	9.3	35.2
Chicago	5.2	20.4	29.4	24.4	Orlando	0.0	1.8	1.5	7.5
Atlanta, Ga.	3.4	28.2	47.9	58.0	Fort Lauderdale	0.0	0.6	0.5	5.2
Other major					West Palm Beach	0.0	1.8	1.0	1.4
centers[a]	19.0	12.6	13.4	16.0	Daytona Beach	0.0	1.2	0.0	0.0
Lesser centers[b]	20.7	6.0	7.2	8.7	Gainesville	0.0	0.6	1.0	1.2
					St. Petersburg	0.0	0.6	0.0	4.2
In-State					Miami Beach	0.0	0.0	0.5	0.7
Jacksonville	50.0	82.8	84.0	66.4	Panama City	0.0	0.0	0.5	0.5
Tampa	10.3	19.8	22.2	23.0	All others	0.0	6.6	14.9	18.1
Tallahassee	3.4	1.8	1.0	6.1					
Pensacola	1.7	7.2	2.6	2.6					

[a] Major centers are identified in Table 12-1.

[b] "Lesser centers" includes all places except major centers.

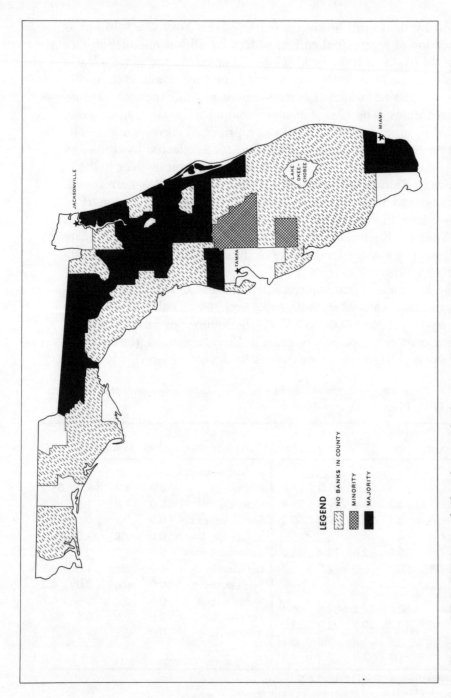

Figure 17-1. Sections of Florida in which a majority of banks chose a principal correspondent in Jacksonville in 1900.

LEGEND

- NO BANKS IN COUNTY
- MINORITY
- MAJORITY

LAKE OKEE-CHOBEE

JACKSONVILLE

TAMPA

MIAMI

the state; and finally, the competition between nationally significant out-state centers and the important cities of Florida.

New York, which slipped slightly in importance between 1900 and 1940 and again between 1950 and 1965, is still a correspondent choice for nearly nine-tenths of all Florida banks. New York's position in Florida, unlike its position in other parts of the South-Atlantic division, has not declined markedly because no major financial center has developed within Florida. The activities of Florida banking centers supplement New York's role within the state.

Another major national banking center, Chicago, and a regional center, Atlanta, now are both much more important in Florida than they were in 1900. Each city had increased its correspondent position by the time of World War II to a level where its banks were selected by at least a fifth of the Florida banks. Atlanta has since moved ahead to a point where its banks serve as correspondents for more than half the Florida institutions. The correspondent roles of other out-state cities declined rather sharply between 1900 and 1940, but stabilized thereafter.

Correspondent banking with Florida cities increased, but the positions of centers did not change in the manner one might expect on the basis of population size alone. Jacksonville expanded its correspondent-banking role to a point where nearly 85 percent of all banks selected a Jacksonville correspondent in 1940 and 1950; by 1965, however, the city was chosen as a correspondent location by only two-thirds of the state's banks. Miami, which was as large a city as Jacksonville in 1940 and even larger in 1950, was chosen only a tenth as often as Jacksonville in 1940 or 1950. Since 1950 there has been an upsurge in the role of Miami banks such that a third of all Florida institutions now maintain a principal correspondent in the center. Tampa rather than Miami was Jacksonville's closest in-state competitor in 1940, even though Tampa was no larger than Miami and was located at a lesser distance from Jacksonville. Although Tampa began to lose ground to Miami in the 1950s, it continues to be selected as a correspondent location by about a fifth of the Florida banks. Thus far, the banking centers in Florida are engaged in a battle for intrastate supremacy, and out-state centers continue to play an important role in linking Florida banking institutions into the national network.

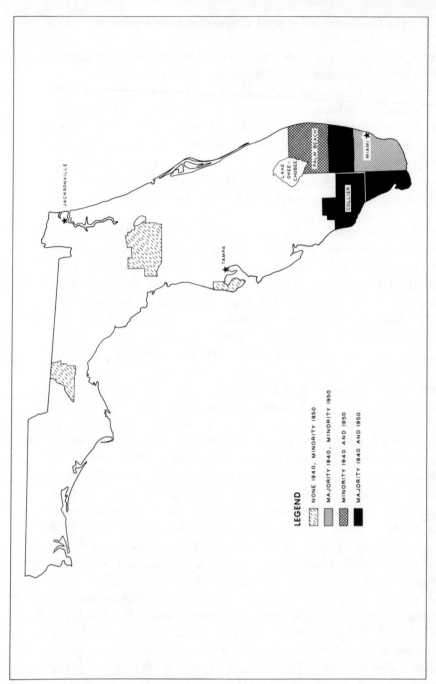

Figure 17-2. Sections of Florida in which a majority of banks chose a principal correspondent in Miami in 1940 and 1950.

Some understanding of the development of a correspondent hinterland can be obtained by examining the competition between Miami and Tampa. It was only in a four-county contiguous area that Miami was selected as a correspondent by at least half the county's banks in 1940. Some banks in Palm Beach county, which adjoined this four-county area, also maintained a Miami correspondent (Figure 17-2). Tampa's hinterland was far more extensive. A majority of the banks in ten counties chose Tampa as a correspondent location, and some banks in an additional four counties used Tampa correspondents (Figure 17-3). Banks in all but five counties in Florida maintained a correspondent tie with a bank in Jacksonville. At least half the banks in two of these counties had correspondents in Tampa as well, and Collier county maintained strong corespondent ties with Tampa and neighboring Miami. No county in the Jacksonville hinterland maintained ties with Miami that did not also maintain ties with Tampa, however, again suggesting that Miami was a weaker rival of Jacksonville than was Tampa.

Tampa's hinterland stretched farther to the south toward Miami than to the north toward Jacksonville. Chicago's position in the financial network was stronger relative to New York's position in the western sections of the nation than in the east according to our analyses of regional differences in correspondent selection patterns. Apparently a "second-ranked" center can compete most effectively with the "first-ranked" center when it occupies an intervening position between the first-ranked center and the contested territory. In the territory which lies between the centers, the competitive advantage lies more clearly with the first-ranked center. The same phenomenon appears in most other delineations of the hinterlands of major centers in the United States. An example is Park and Newcomb's display of newspaper circulation which shows the hinterland for Minneapolis papers extending much further to the west than to the east toward Chicago (Park, 1952: 219). The phenomenon also is suggested by the business-loan analyses reported by Duncan et al. (1960: 120).

Despite Miami's significant population growth in the 1940s, its correspondent hinterland in 1950 did not differ substantially from its 1940 hinterland. Some banks in three distant counties had chosen to maintain correspondents in Miami, but there was no expansion of the contiguous hinterland. Tampa's network of correspondent ties

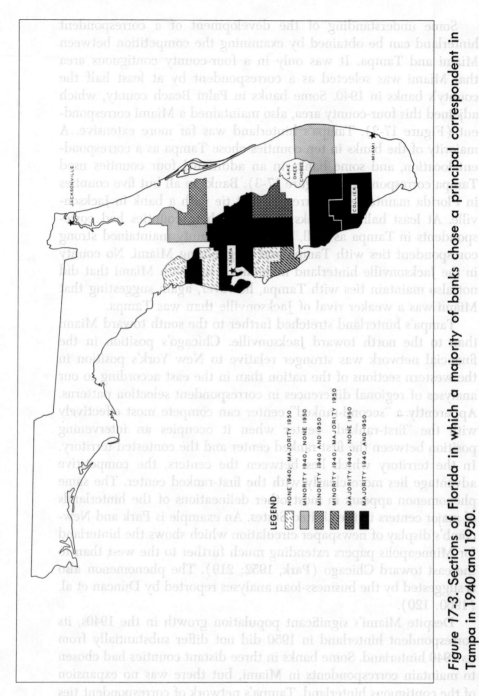

Figure 17-3. Sections of Florida in which a majority of banks chose a principal correspondent in Tampa in 1940 and 1950.

had not been altered greatly either. Although a loss of business occurred in Palm Beach county, its role in other counties near Miami continued to be strong; and some consolidation of its hinterland was evident. Jacksonville's primacy remained unchallenged by Miami and Tampa. Indeed, the number of counties in which no bank maintained a principal correspondent in Jacksonville fell from five to three.

There was no suggestion during the 1940s of the rapid expansion of Miami's hinterland that would take place in the next fifteen years. Miami's role as a correspondent for banks along the eastern coast of Florida increased enormously after 1950. Although its position is weaker in the western section of the peninsula where Tampa had carved out a hinterland, some banks in a number of western counties now select Miami (Figure 17-4). Tampa has not exhibited the same growth in correspondent ties as Miami, but it has not fared badly. The core of Tampa's 1950 hinterland remained intact in 1965, and there now is an outer ring of counties bordering the Atlantic in which some banks select Tampa (Figure 17-5).

Jacksonville's position within the state has begun to deteriorate although it is still stronger than its rivals of Miami and Tampa. In five counties, no bank selected a Jacksonville correspondent in 1965. Three of these, Dade, Broward, and Collier, are in Miami's hinterland; the remaining two, Hillsborough and Manatee, are in Tampa's domain. Okeechobee and Indian River counties, two of the five counties in which some banks, but fewer than half, choose Jacksonville as a correspondent location, now have established strong correspondent ties to Miami. Although Miami banks now have demonstrated their ability to displace, rather than merely supplement, the correspondent services provided by Jacksonville, more often the correspondent role is shared.

Miami's rise as a correspondent center may be hindered by organizational factors. Not only is Jacksonville designated a "Branch" city; several of Jacksonville's major banks maintain affiliate banks throughout the state that are quasi-chains. In point of fact, members of banking groups headquartered in Jacksonville were no less likely in 1965 to maintain correspondents in Miami or Tampa than were independent banks in the same locality. The quasi-chain organization does strengthen the correspondent-banking position of Jacksonville, however, because the affiliate banks are more likely

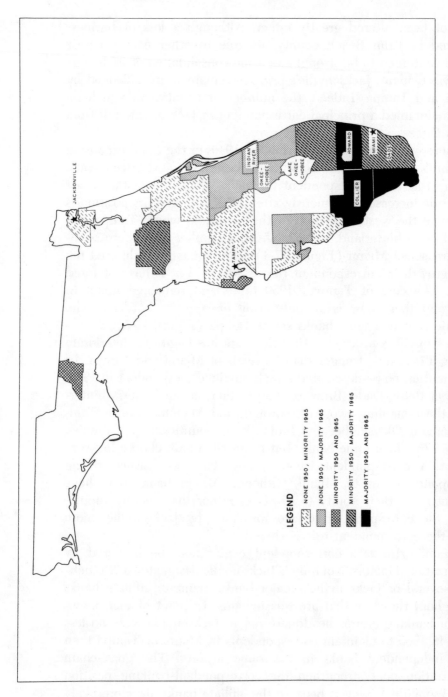

Figure 17-4. Sections of Florida in which a majority of banks chose a principal correspondent in Miami in 1950 and 1965.

LEGEND

NONE 1950, MINORITY 1965

NONE 1950, MAJORITY 1965

MINORITY 1950 AND 1965

MINORITY 1950, MAJORITY 1965

MAJORITY 1950 AND 1965

JACKSONVILLE

INDIAN RIVER

OKEE-CHOBEE

LAKE OKEE-CHOBEE

COLLIER

HOWARD

MIAMI

TAMPA

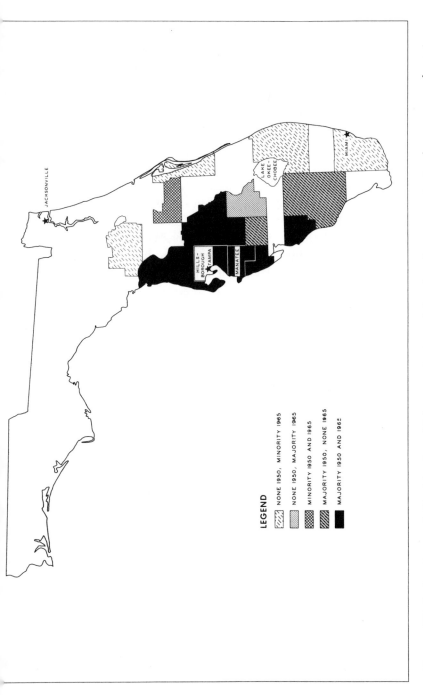

Figure 17-5. Sections of Florida in which a majority of banks chose a principal correspondent in Tampa in 1950 and 1965.

LEGEND

- NONE 1950, MINORITY 1965
- NONE 1950, MAJORITY 1965
- MINORITY 1950 AND 1965
- MAJORITY 1950, NONE 1965
- MAJORITY 1950 AND 1965

to have correspondents in Jacksonville than are other banks in the same county. (If the relative frequency with which chain banks select a given correspondent location were identical with the frequency of selection on the part of nonchain banks in the same county, the number of banks selecting Miami would be nine and the numbers selecting Tampa and Jacksonville would be three and thirty, respectively. The actual numbers of chain banks selecting these centers are: Miami, nine; Tampa, four, and Jacksonville, thirty-nine.)

Jacksonville, moreover, lies more nearly in the path of north-south commerce, thereby enjoying a locational advantage over Tampa and Miami. Miami is particularly disadvantaged. Located on the southern tip of the peninsula, Miami has virtually no contiguous area to organize as a hinterland to the south, little to the west, and none to the east. Since nearly all of Florida lies to the north, a flow of checks into Miami would involve a certain amount of backtracking for exchanges with almost all sections of the nation.

Miami is disadvantaged not only by geographic factors, but also by the basic source of its growth, namely, tourism and retirement. Miami has been slow in developing the kinds of hinterland activities that nurture banking strength. Tampa and Jacksonville, although enjoying a tourist industry, developed commercial services oriented to the areas around them, acting as wholesaling and collection centers. (The activities of the respective centers have been described by Duncan et al. [1960: 528–531, 437–442, and 537–543].) Miami's economy has been less closely linked to a hinterland in Florida, and thus its growth as a correspondent-banking center has not paralleled its expansion in population. In this regard, Miami provides an apt illustration of Gras' point that while a metropolis requires a large population, this is itself an insufficient precondition for metropolitan status (1922: 294).

Chapter 18

CORRESPONDENCE ABOUT CORRESPONDENTS

A number of the large banks in the cities studied did not list their principal correspondents in the 1965 *Rand McNally Bankers Directory* although they had done so in earlier periods. It is now a common practice for big banks to indicate only that they have correspondents in the principal cities of the United States. Although this is undoubtedly true, it was necessary to write to them in order to learn the cities in which their principal correspondents were located. In passing, the banks also were asked why they chose to omit listing their major correspondents. The answers were rather surprising; although a large number of banks provided their principal correspondents on request, a fair number declined to provide this information.

A follow-up letter was sent indicating that we needed to know only the cities in which the principal correspondents were located and, in addition, pointing out that the data would be aggregated in such a way that anonymity was assured. Given this assurance, the vast majority of the banks contacted provided the necessary information. (If a bank failed to specify its principal correspondents, their locations were "guessed" on the basis of known correspondents of banks in that city. Such "guesses" are relatively few, but no information about their distribution is provided here to ensure the anonymity of respondents.) From these rounds of correspondence came some indications of how bankers perceive the roles of their banks and home communities in the financial network.

The possibility of inadvertent disclosure of identity precludes any systematic treatment of the responses. Nonetheless, it is possible to convey some impression of the bankers' views. First, New York and its large financial institutions play a unique role in the banking network. Second, small banks in major centers and banks located outside these communities are distinctly subordinate to or non-competitive with the large banks in the large cities. Third, relations between sizable banks in the nation's largest communities, save New York, are ambiguous; each seeks an ill-defined "fair share" of its correspondent's business (reciprocity) and a "lion's share" of the business generated by other banks (competition).

Several leading New York City banks remained unwilling to list their principal correspondents. Their reasons for refusing included testimony to the city's special role in the national banking system and offer further evidence of New York's primary position in providing correspondent services for banks located elsewhere in the nation, including those in other major centers. The excerpts which follow indicate the one-sided nature of bank-correspondent selection so far as three leading New York institutions are concerned:

> New York City, being the financial capital of the country, is in the position of rendering services of various sorts to banks throughout the country but, in general, these other banks do not offer services to New York City banks except in the area of the collection of cash items drawn on banks in their locality.
>
> While we have several collection accounts of this nature, these represent special situations in which the other institution can give us quicker service than the Federal Reserve Bank through which most of our collections are effected.
>
> We do not feel that it serves any purpose for us to list the relatively few collection accounts which we maintain with domestic banks, which is what I believe you have in mind. The main correspondent relations insofar as balances are concerned run from other banks to us.

> We do not maintain balances with other banks except in rare instances or special circumstances. A list of these would in no way indicate our principal relationships. These are for the specific purpose of

collecting checks or other items and really are not referred to as correspondent accounts.

. . . I am sorry I cannot give you exactly what you asked for. This is not the result of reluctance on our part to divulge any information, but since the relationship is from the other side.

Balances at one or more money centers are near-essential for country banks for reasons of varying importance. . . . Large city banks perform many functions, and we offer our services to such banks as may wish them and presently serve almost all the larger banks in the country.

Responses from bankers located in some other cities suggest, on the other hand, that their reluctance to list principal correspondents reflects, in part, an ambiguity about the positions of their bank in the financial hierarchy. In fairness, many large banks outside New York indicated that any knowledgeable person would realize that an institution of such size would have correspondents virtually everywhere and, further, that the space necessary for listing them would be both costly and better used for other purposes. The following two quotations, from a west coast banker and an official of an Ohio institution, respectively, are representative of a number of responses:

We do not attempt to list correspondent banks due to space limitation in the Directory and to the fact that in the trade it is generally understood that a bank the size of ours will have correspondents in all of the major cities throughout the United States.

. . . simply because of the space involved which . . . can be used more productively with other information. We feel certain that most people who might come to us for help in a distant city will come whether we list a correspondent in that city or not and we feel sure we can accomplish their mission regardless. I am sure that in some instances anonymity may be desirable but these would be the rare exception.

That the failure to list principal correspondents sometimes is due to more than the limitations and cost of directory space, however, is suggested by the fact that a number of banks were willing to provide the information requested only after they were assured of anonymity or with the understanding that not the names of banks serving as correspondents, but only their locations were needed.

Why should banks be wary of providing a list of their principal correspondents? Reciprocity and competition in interbank relations appear to be at stake, especially when the banks are relatively large and it is not clear which city is the more important center in the national banking network. Seldom will there be full reciprocity between banks. Often a bank will serve as a correspondent for several banks located in another center and, in turn, select as a correspondent several banks located in that center. Although any bank can easily compare how much business it receives from another bank with the amount it provides for that bank, the total amount of business that local competitors are receiving from the same correspondent is not always known. Were one of the bank's customers to realize that it was not drawing its proportionate share of the business directed by its correspondent to the city, the interbank relation would be strained; and the institution failing to receive reciprocal business could apply pressure for more business through the deposits it maintains with the other bank. Excerpts from the letters of three bankers tend to support this line of reasoning:

> If we were to designate a selected group of these banks as being our principal correspondents, we would be most certain to offend some of the banks who were not given that designation. I'm sure you are aware that when such problems arise it is most difficult to maintain a friendly relationship that may have taken so much time, effort and expense to develop.

> Most large banks have more than one correspondent in a large city, but to just indicate the one which is deemed to be the most important would not be wise. In almost every case, if two banks in a given city had a deposit from us, and if each was rather substantial, it could be fairly well assumed that each bank would think it was the principal correspondent.

> If the report disclosed we maintained even one correspondent in, say, Los Angeles, although that correspondent bank remained unnamed, we would have all the others on our back, obviously knowing that they were not the favored one account.

The correspondent-bank selections of small banks in major financial centers or of banks in lesser centers seemingly need less confidential treatment. These banks are clearly subordinate to the large financial institutions in major centers which serve as their correspondents (Lieberson, 1961). There is no question of reciprocity or competition. In return for the placement of noninterest-bearing balances with their correspondents, the small institutions and banks outside major financial centers receive certain services. As one banker wrote,

> Small banks, on the other hand, like to list their city correspondents because this is a way of saying that the services of big banks like the Chase, First National City, Bankers Trust and so forth are, in effect, available to their customers.

As one moves up the organizational hierarchy to large banking institutions in major financial centers, correspondent relations become more complicated. A large bank may have more than one correspondent in a given center. Moreover, the bank wishes to serve as a correspondent for as many of that center's banks as possible and to receive their deposits. It would hardly do for a Kansas City bank with important deposits from, say, four different Baltimore banks, for example, to reveal that a single institution receives most of its Baltimore business. In other words, in the upper echelons where there is no clear-cut ranking, the big-bank correspondents in a city need not assume that the flow will be all in one direction. An officer in charge of correspondent banking put it this way:

> We have very good reasons for not listing our principal correspondents in the bank directory. No. 1, we don't want our competitors to know which bank we consider our favorite in the major centers, and No. 2, we don't want the banks in New York and Chicago, for example, to know which bank in their city we consider our favorite or principal correspondent account. In New York we carry accounts with ten banks, and I am sure that two or three of them think that they are our principal correspondent.

The competitive thrust also is illustrated in several other letters:

> The primary reason . . . is that we have reciprocal balances of varying sizes and we would prefer not to name some banks as principal correspondents at the risk of offending others who might maintain profitable accounts with us.
>
> . . . the listing of such information may on the one hand be helpful when read by the friends of the correspondent listed, but at the same time be harmful when other banks are omitted.

We do not list our principal correspondents . . . because in certain cities we have some rather delicate relationships and would not want to leave certain names out.

In many instances we do have more than one relationship in a given city, state, or region. Some naturally are larger or older or more active, but we have never made any such designations as principal and secondary correspondents. We certainly would not want any of our customers to feel they are getting less than our best service or ultimately we would end up with only a single correspondent, our principal one.

Not only is the relation between a pair of large banks located in two relatively important banking centers ambiguous; the banks may, in fact, be in competition for prospective correspondent and business accounts from other parties. A bank which is seeking to expand its correspondent business is, in the words of one banker, "actively competing with many, if not all, of the banks that would be listed."

Several banks indicated that certain confidential relations existed which made it desirable not to report their principal correspondents. "We are hesitant to name . . . *all* or *some* of our large city correspondents. Many considerations—including services on behalf of corporate customers—are involved in these relationships and we have therefore made a point of not listing these banks," wrote the public relations director of a bank. Another bank indicated that the "banks and bankers with whom we have a working relationship perform more or less specialized services for us which could be and often are confidential." Some banks maintained that it was impossible to single out certain correspondents as their principal ones on the grounds that all served important, though distinctive functions.

The big banks' current practice of omitting their principal correspondents in directory advertisements seems to follow, in part, from the absence of a clearly defined hierarchy among the large banks located in major centers outside the money market of New York. This is consistent with our contention that the correspondent-banking relations between major centers are far more diffuse and less ordered now than at the beginning of the century, when nearly all banks identified publicly their principal correspondents.

APPENDIX

APPENDIX

Appendix Table A. Population (in thousands) of Centers with Populations Greater than or Equal to That of New Orleans in Given Census Year, 1820 to 1960

Center	1820	1830	1840	1850	1860	1870	1880	1890	1900[a]	1910[b]	1920[b]	1930[b]	1940[c]	1950[d]	1960[d]
New York	124	203	313	516	814	1,362[e]	1,805[e]	2,354[e]	4,023	5,621	6,765	9,423	11,150	12,296	14,115
Philadelphia	82[f]	132[f]	206[f]	368[f]	566	674	847	1,047	1,458	1,761	2,116	2,399	2,574	2,922	3,635
Baltimore	63	81	102	169	212	267	345[g]	445[g]	543	609	761	877	946	1,162	1,419
Boston	43	61	93	137	193[h]	325[i]	420[i]	571[i]	905	1,096	1,260	1,545	2,052	2,233	2,413
New Orleans	27	46	99[j]	116	169	191	216	242	291	344	392	477	536	660	845
St. Louis	—	—	—	—	(161)	311	351	452	612	758	863	1,094	1,201	1,401	1,668
Chicago	—	—	—	—	(109)	299	503	1,100	1,768	2,316	2,940	3,870	4,364	4,921	5,959
Cincinnati	—	—	—	—	(161)	216	255	297	414	464	504	605	708	813	994
Pittsburgh	—	—	—	—	—	(139)[k]	235[k]	344[k]	622	788	898	1,312	1,444	1,533	1,804
San Francisco	—	—	—	—	—	(149)	234	299	444	627	807	1,104	1,318	2,022	2,431
Minneapolis	—	—	—	—	—	—	(88)[l]	298[l]	369	521	622	784	838	987	1,377
Cleveland	—	—	—	—	—	—	(160)	261	402	587	861	1,048	1,211	1,384	1,785
Buffalo	—	—	—	—	—	—	(155)	256	373	456	555	735	792	896	1,054
Milwaukee	—	—	—	—	—	—	—	(204)	305	401	497	660	728	829	1,150
Detroit	—	—	—	—	—	—	—	(206)	305	484	1,080	1,837	2,186	2,752	3,538
Washington	—	—	—	—	—	—	—	(230)	292	349	473	554	859	1,287	1,808

Appendix Table A. (continued)

Center	1820	1830	1840	1850	1860	1870	1880	1890	1900[a]	1910[b]	1920[b]	1930[b]	1940[c]	1950[d]	1960[d]
Los Angeles	—	—	—	—	—	—	—	—	(114)	379	728	1,778	2,722	3,997	6,489
Kansas City	—	—	—	—	—	—	—	—	—	(336)	451	565	590	698	921
Houston	—	—	—	—	—	—	—	—	—	—	—	—	(456)	701	1,140
Dallas	—	—	—	—	—	—	—	—	—	—	—	—	—	(539)	932
Seattle	—	—	—	—	—	—	—	—	—	—	—	—	—	(622)	864
Miami	—	—	—	—	—	—	—	—	—	—	—	—	—	(459)	853

( ). Population at beginning of decade during which center's population surpassed that of New Orleans in size.

a. Population of central city or cities of 1910 metropolitan district plus half population in adjacent area of district.

b. Population of central city or cities of metropolitan district of that year plus half population in adjacent area of district.

c. Population of 1940 metropolitan district adjusted by ratio of 1950 population of urbanized area to 1950 population of 1940 district.

d. Population of urbanized area of that year.

e. Population of Kings County.

f. .6, .7, .8, .9 of Philadelphia County at successive dates.

g. .5 population of territory to be annexed.

h. Suffolk County.

i. Suffolk County plus .2 Middlesex County.

j. Estimated from city/parish ratios, 1830 and 1850, and 1840 parish population.

k. Includes Allegheny city.

l. Includes St. Paul city.

Source: Twelfth Census (1900), Vol. 1, Population, Part 1, Counties, Table 4, and Cities, Towns, Villages, and Boroughs, Table 6; Bureau of the Census, *Population: The Growth of Metropolitan Districts in the United States: 1900–1940*, by W. S. Thompson (Washington: Government Printing Office, 1948), Table 3; 1950 Census of Population, Vol. 1, United States Summary, Table 30; and 1960 Census of Population, Vol. 1, United States Summary, Table 22.

BIBLIOGRAPHY

ALDERFER, E. B. and H. E. MICHL (1957) *Economics of American Industry.* New York: McGraw-Hill.

BERRY, B. J. L. (1965) "Research frontiers in urban geography," in P. M. Hauser and L. F. Schnore (eds.) *The Study of Urbanization.* New York: John Wiley.

Board of Governors of the Federal Reserve System (1943) *Banking and Monetary Statistics.* Washington, D.C.: National Capital Press.

— (1938) "The history of reserve requirements for banks in the United States." *Federal Reserve Bulletin* 24: 953-72.

BOND, L. H. (n.d.) Personal communication.

BUER, M. C. (1926) *Health, Wealth, and Population in the Early Days of the Industrial Revolution.* London: George Routledge.

DERRY, T. K. and T. I. WILLIAMS (1961) *A Short History of Technology.* New York: Oxford University Press.

DUNCAN, O. D. (1966) "Path analysis: sociological examples." *American Journal of Sociology* 72: 1-16.

—, W. R. SCOTT, S. LIEBERSON, B. DUNCAN, and H. H. WINSBOROUGH (1960) *Metropolis and Region.* Baltimore: Johns Hopkins Press.

FAULKNER, H. U. (1943) *American Economic History.* New York: Harper.

FINNEY, K. (1958) *Interbank Deposits.* New York: Columbia University Press.

GOLDSMITH, R. W. (1958) *Financial Intermediaries in the American Economy Since 1900.* Princeton, N.J.: Princeton University Press.

GRAS, N. S. B. (1922) *An Introduction to Economic History.* New York: Harper.

HARRIS, C. D. and E. L. ULLMAN (1945) "The nature of cities." *Annals of the American Academy of Political and Social Science* 242: 7–17.

JOHNSON, E. R. and T. W. VAN METRE (1921) *Principles of Railroad Transportation.* New York: D. Appleton.

LAMPARD, E. E. (1965) "Historical aspects of urbanization," in P. M. Hauser and L. F. Schnore (eds.) *The Study of Urbanization.* New York: John Wiley.

— (1955) "The history of cities in the economically advanced areas." *Economic Development and Cultural Change* 3: 81–136.

LIEBERSON, S. (1961) "The division of labor in banking." *American Journal of Sociology* 66: 491–96.

—, and K. P. SCHWIRIAN (1962) "Banking functions as an index of inter-city relations." *Journal of Regional Science* 4: 69–81.

McKENZIE, R. D. (1933) *The Metropolitan Community.* New York: McGraw-Hill.

McNEILL, W. H. (1963) *The Rise of the West.* Chicago: University of Chicago Press.

MADDEN, C. H. (1959) *The Money Side of "The Street."* New York: Federal Reserve Bank of New York.

MILLER, S., Jr. (1950) "History of the modern highway in the United States," in J. Labatut and W. J. Lane (eds.) *Highways.* Princeton, N.J.: Princeton University Press.

OGBURN, W. F. (1937) "National policy and technology," in *Technological Trends and National Policy.* Washington, D.C.: Government Printing Office.

PARK, R. E. (1952) *Human Communities.* New York: Free Press.

PERLOFF, H. S., E. S. DUNN, Jr., E. E. LAMPARD, and R. F. MUTH (1960) *Regions, Resources, and Economic Growth.* Baltimore: Johns Hopkins Press.

POTTER, A. A. and M. M. SAMUELS (1937) "Power," in *Technological Trends and National Policy.* Washington, D.C.: Government Printing Office.

ROBERTSON, R. M. (1954) "St. Louis: central reserve city, 1887–1922." *Monthly Review of the Federal Reserve Bank of St. Louis* 36: 85–92.

ROBINS, S. M. and N. E. TERLECKYJ, with the collaboration of I. O. SCOTT, Jr. (1960) *Money Metropolis: A Locational Study of Financial Activity in the New York Region.* Cambridge, Mass.: Harvard University Press.

ROBINSON, R. I. (1960) *Postwar Market for State and Local Government Securities.* Princeton, N. J.: Princeton University Press.

RODKEY, R. G. (1934) "Legal reserves in American banking." *Michigan Business Studies* 6: 363–483.

ROSE, A. C. (1950) "The highway from the railroad to the automobile," in J. Labatut and W. J. Lane (eds.) *Highways*. Princeton, N.J.: Princeton University Press.

SHIMKIN, D. B. (1952) "Industrialization, a challenging problem for cultural anthropology." *Southwestern Journal of Anthropology* 8: 84–91.

Standard & Poor's *Security Dealers of North America* (1964) New York.

TAYLOR, C. T. (1957) "Southern vs. non-Southern underwriting of municipals." *Southern Economic Journal* 24: 158–69.

THOMPSON, W. R. (1965) "Urban economic growth and development in a national system of cities," in P. M. Hauser and L. F. Schnore (eds.) *The Study of Urbanization*. New York: John Wiley.

United States Department of Commerce, Bureau of the Census. *Twelfth Census of the United States: 1900—Population,* Vol. 2, Pt. 2. *Manufactures,* Vol. 7, Pt. 1, Vol. 8, Pt. 2, Vol. 10, Pt. 4.

—*Thirteenth Census of the United States: 1910—Population,* Vol. 4. *Manufactures,* Vol. 8, Vol. 9.

—*Census of Electrical Industries: 1917—Electric Railways.*

—*Fourteenth Census of the Uniter States: 1920—Population,* Vol. 4. *Manufactures,* Vol. 9, Vol. 10.

—*Sixteenth Census of the United States: 1940—Manufactures, 1939,* Vol. 1, Vol. 3. *Census of Business, 1939,* Vol. 2, Vol. 3.

—*1947 Census of Manufactures,* Vol. 3.

—*1948 Census of Business,* Vol. 5, Vol. 7.

—*1954 Census of Business,* Vol. 4, Vol. 6; *1954 Census of Manufactures,* Vol. 3.

—*1958 Census of Business,* Vol. 4, Vol. 6; *1958 Census of Manufactures,* Vol. 3.

—*1960 Census of Population,* Vol. 1, State Reports.

—*1963 Census of Business,* Vol. 5, Vol. 7; *1963 Census of Manufactures,* Vol. 3; *1963 Census of Transportation,* Vol. 3.

—*Current Population Reports,* Series P–25, No. 347.

—*Historical Statistics of the United States* (1960); *Continuation to 1962 and Revisions* (1965).

United States Senate (1914) *Location of Reserve Districts in the United States.* Document 485, Senate Documents, Volume 16, 63rd Congress, 2nd Session. Washington, D.C.: Government Printing Office.

VERNON, R. (1963) *Metropolis 1985.* Garden City, N. Y.: Doubleday.

WADE, R. C. (1959) *The Urban Frontier.* Cambridge, Mass.: Harvard University Press.

WILSON, G. L. (1942) "Petroleum pipe-line transportation," in *Trans-*

portation and National Policy. Washington, D.C.: Government Printing Office.

WINSBOROUGH, H. H., W. R. FARLEY, and N. D. CROWDER (1966) "Inferring an Hierarchy from Flow Data." Paper presented at the annual meeting of the American Sociological Association.

INDEX

295

PLACES

Passaic 80
Paterson 80
Pensacola 270
Peoria 183
Philadelphia 44, 46, 47, 51, 80, 114
    financial function 101, 103, 108,
        110, 112, 114, 115, 201, 210,
        211, 213, 215, 216
    industrial structure 22, 62, 72, 81,
        82, 84, 86, 87, 91, 112, 115, 123,
        136, 147, 150, 155, 163, 165,
        175, 218, 221, 222, 245, 250,
        252
    regional relations 47, 51, 59, 74,
        108, 110, 193, 206, 227, 229,
        230, 231, 237
    timing of growth 22, 37, 39, 51,
        53, 55, 95, 97, 98, 148
Phoenix 23, 24, 151, 176, 185, 209
Pittsburgh 23, 73, 76, 80, 95, 107,
    112, 114, 115, 184, 213
    industrial structure 22, 58, 71, 72,
        84, 86, 87, 88, 91, 112, 115, 122,
        136, 147, 155, 159, 163, 165,
        167, 218, 221, 245, 248, 252
    regional relations 56, 71, 74, 163,
        206, 229, 230, 236, 237
Portland 103, 105
Providence 102, 103
Pueblo 105
Quincy 80
Richmond 151, 183
Rochester 101, 103
Salem 80
Salt Lake City 102, 105, 209
San Antonio 102, 105, 254, 255, 257
San Bernardino 23, 24, 141, 151, 176
San Diego 23, 24, 151, 172, 184, 185
San Francisco 23, 55, 56, 71, 80, 127,
    130, 151, 205, 217, 221
    financial function 114, 184, 204,
        210, 213, 215
    industrial structure 22, 24, 72, 84,
        93, 115, 138, 147, 150, 170, 174,
        175, 246, 248, 250, 251, 252,
        253

regional relations 22, 24, 56, 85,
    108, 110, 111, 162, 170, 194,
    196, 206, 209, 227, 230, 237
San Jose 23, 24, 151, 176
Savannah 103
Seattle 23, 105, 106, 179, 184, 215
    industrial structure 24, 136, 150,
        152, 172, 174, 248, 252
    regional relations 24, 194, 198, 227,
        230, 237
Sioux City 102, 105
Sioux Falls 102
Somerville 80
South Omaha 105, 107
Spokane 105, 106
St. Joseph 102, 103
St. Louis 55, 56, 71, 95, 102, 184,
    199, 254
    financial function 101, 104, 108,
        112, 114, 182, 183, 201, 203,
        210, 213, 215, 216
    industrial structure 22, 62, 72, 84,
        86, 91, 112, 115, 147, 150, 155,
        161, 165, 170, 176, 218, 221,
        222, 245, 248, 252
    regional relations 22, 55, 108, 110,
        194, 196, 206, 208, 229, 230,
        231, 236, 237, 255, 257, 261
St. Paul 80, 103
St. Petersburg 270
Tacoma 105, 106, 184
Tampa 270, 271, 273, 275, 277, 280
Toledo 183
Topeka 105
Tucson 141, 209
Tulsa 106, 183
Waco 105
Waltham 80
Washington 23, 107, 147, 196, 205,
    210, 213
    industrial structure 22, 76, 82, 86,
        115, 131, 136, 145, 150, 163,
        170, 248, 249, 252
West Hoboken 80
Wichita 105
Youngstown 184

SUBJECTS

ABOUT THE AUTHORS

BEVERLY DUNCAN is a Research Associate of the Population Studies Center at the University of Michigan. She currently is editor of *Demography*, a quarterly journal of the Population Association of America. Among her publications are *The Negro Population of Chicago* (University of Chicago Press, 1957, with O. D. Duncan) and *Housing a Metropolis* (Free Press of Glencoe, 1960, with P. M. Hauser).

STANLEY LIEBERSON is Professor of Sociology at the University of Washington and Director of the Center for Studies in Demography and Ecology. Among his publications are *Language and Ethnic Relations in Canada* (John Wiley, in press), *Ethnic Patterns in American Cities* (Free Press of Glencoe, 1963), and editor, *Explorations in Sociolinguistics* (Indiana University, 1967).

Both Beverly Duncan and Stanley Lieberson were among the co-authors of *Metropolis and Region* (Johns Hopkins, 1960).

DATE DUE

SEP 24 '78			
GAYLORD			PRINTED IN U.S.A